PHILOSOPHY & THE FUTURE

A TRANSHUMANIST EXAMINATION OF WHERE WE'RE GOING

ZOLTAN ISTVAN

Copyright (c) 2020 Rudi Ventures LLC
(otherwise, all permissions granted for use)
Published by Rudi Ventures LLC
Cover Design: Rachel Edler
ISBN#: 978-1-7363426-1-9

****

AUTHOR'S NOTE

While these essays have been arranged and edited for readability, many of them appear similar (if they are not new) to how they were originally published. Attempts have been made to preserve the context and moment in time they were written. Some articles contain British spelling. Publishing information and fact checks can be found by utilizing the Appendix.

****

TABLE OF CONTENTS

INTRODUCTION

CHAPTERS

I: Exploring the Future

1) Post-Earther: Nature Isn't Sacred and We Should Replace It

2) Silicon Valley is Ditching Pascal's Wager: New Ideas like Quantum Archaeology are Trying to Challenge Religion and Even the Permanence of Death

3) Programming Hate into AI will be Controversial, but Probably Necessary

4) Future of Transhumanist Tech may Soon Change the Definition of Disability

5) What if One Country Achieves the Singularity First?

6) Why I'm Debating an Anarcho-Primitivist Philosopher About the Future

7) Do We Have Free Will Because God Killed Itself?

8) Will Capitalism Survive the Coming Robot Revolution?

9) The Culture of Transhumanism is About Self-Improvement

II: Early Writings

10) When Does Hindering Life Extension Science Become a Crime?

11) Cryonics, Special Needs People, and the Coming Transhumanist Future

12) Despite Skepticism, Many People May Embrace Radical Transhumanist Technology in a Futurization of Values

13) Origami Cranes: Who is Responsible for this Child's Death? (Introduction to the World's First Mainstream Media Column on Transhumanism: *Psychology Today's: The Transhumanist Philosopher*)

14) Transhumanists Frown on Talk of Genetic Engineering Moratorium

15) A World Future Society Conference Speech: Everyone Faces a Transhumanist Wager

III: Personal Notes

16) Should I Have had my Cat Cryonically Preserved?

17) The New American Dream: Let the Robots Take our Jobs

18) Baggage Culture and Why Embracing Transhumanism Doesn't Come Easy

19) Should Surfing be Allowed During the Pandemic?

20) Trojan Horse: Why I'm Running for President as a Republican

21) A Letter About Coronavirus, the Longevity Movement, & Why Quarantining is Killing Us

22) Death Threats, Freedom, Transhumanism, and the Future

IV: Cultural Philosophy

23) How Brain Implants (and Other Technology) Could Make the Death Penalty Obsolete

TABLE OF CONTENTS

INTRODUCTION

CHAPTERS

I: Exploring the Future

1) Post-Earther: Nature Isn't Sacred and We Should Replace It

2) Silicon Valley is Ditching Pascal's Wager: New Ideas like Quantum Archaeology are Trying to Challenge Religion and Even the Permanence of Death

3) Programming Hate into AI will be Controversial, but Probably Necessary

4) Future of Transhumanist Tech may Soon Change the Definition of Disability

5) What if One Country Achieves the Singularity First?

6) Why I'm Debating an Anarcho-Primitivist Philosopher About the Future

7) Do We Have Free Will Because God Killed Itself?

8) Will Capitalism Survive the Coming Robot Revolution?

9) The Culture of Transhumanism is About Self-Improvement

II: Early Writings

10) When Does Hindering Life Extension Science Become a Crime?

11) Cryonics, Special Needs People, and the Coming Transhumanist Future

12) Despite Skepticism, Many People May Embrace Radical Transhumanist Technology in a Futurization of Values

13) Origami Cranes: Who is Responsible for this Child's Death? (Introduction to the World's First Mainstream Media Column on Transhumanism: *Psychology Today's: The Transhumanist Philosopher*)

14) Transhumanists Frown on Talk of Genetic Engineering Moratorium

15) A World Future Society Conference Speech: Everyone Faces a Transhumanist Wager

III: Personal Notes

16) Should I Have had my Cat Cryonically Preserved?

17) The New American Dream: Let the Robots Take our Jobs

18) Baggage Culture and Why Embracing Transhumanism Doesn't Come Easy

19) Should Surfing be Allowed During the Pandemic?

20) Trojan Horse: Why I'm Running for President as a Republican

21) A Letter About Coronavirus, the Longevity Movement, & Why Quarantining is Killing Us

22) Death Threats, Freedom, Transhumanism, and the Future

IV: Cultural Philosophy

23) How Brain Implants (and Other Technology) Could Make the Death Penalty Obsolete

24) Marriage Won't Make Sense When We Live 1000 Years

25) Do We Really Hate Trump and Clinton So Much?

26) Is It Time to Consider Restricting Human Breeding?

27) How Transhumanist Tech Will Correct Reality's Typos: AR, VR, and Brain Implants Will be our Editors

V: Artificial Intelligence and Self

28) Why I Advocate for Becoming a Machine

29) If Our Thoughts Live Forever, Do We Too?

30) An AI Global Arms Race is Looming

31) Is an Affair in Virtual Reality Still Cheating?

32) The Morality of Artificial Intelligence and the Three Laws of Transhumanism

33) When Computers Insist They're Alive

VI: Economic Ideas

34) Capitalism 2.0: The Economy of the Future Will be Powered by Neural Prosthetics

35) Technology Will Replace the Need for Big Government

36) Facing Up to Facial Recognition, and Why We Should Embrace It

37) In 15 Years We'll be Able to Upload Education to our Brains. So Can I Stop Saving for my Kids' College?

38) Delayed Fertility Advantage: Transhumanist Science will Free Women from their Biological Clocks

VII: Atheism Vs. Religion

39) Mind Uploading Will Replace God

40) Some Atheists and Transhumanists are Asking: Should it be Illegal to Indoctrinate Kids with Religion?

41) Christian Relativism: Upgrading Religion for the 21st Century & Why Christianity is Forcibly Evolving to Cope with Science and Progress

42) I Visited a Community Where People Upload Their Personalities to 'Mindfiles' so They can Live on After Death

43) Will the Religious Try to Convert Superintelligence When it Arrives?

VIII: Political Navigation

44) The Transhumanist Party's Founder on the Future of Politics

45) Revolutionary Politics are Necessary for Transhumanism to Succeed

46) The Cyborgs of the Future—and of Today—Need the Transhumanist Bill of Rights

47) Expanding the Non-Aggression Principle: The Future of Libertarianism Could be Radically Different

48) Avoid War and Don't Get Complacent About Freedom in America

IX: Down the Rabbit's Hole

49) The Coming Genetic Age of Humans Won't be Easy to Stomach

50) The Privacy Enigma: Liberty Might be Better Served by Doing Away with Privacy

51) The Next Step for Veganism Is Ditching Our Bodies and Digitizing Our Minds

52) Augmentation, Hivemind & the Omnipotism

53) Should I Have let my Daughter Marry our Robot?

54) Why I'm Running for President—and Got a Chip Implanted in my Hand: Maybe the Difference Between RFID and LSD is just Another Door of Perception

X: The Trilogy

55) The Internet Will Become Self Aware When Aliens Wake it Up

56) Why Haven't We Met Aliens Yet? Because They've Evolved into AI

57) The Language of Aliens Will Always be Indecipherable

APPENDIX

AUTHOR'S BIOGRAPHY

ABOUT THE BOOK

******

INTRODUCTION

I studied Philosophy in college, and it gives me much joy to publish my own book of philosophical essays. My college senior thesis was on "brains in a vat." I attempted to convince my professor via various arguments that humans could distinguish themselves from being just manipulated brains in some aliens' aquariums versus actual living, thinking mammals here on Planet Earth. I don't think I was very successful, which goes to show the slippery nature of epistemology, language, and philosophy.

That said, the 50+ essays below are simpler, less technical philosophy. Some might even call a portion of these writings: pop philosophy. However, I see them as discourses from a curious mind that aims to tackle the big questions and issues of where humanity is going.

Throughout the span of my public figure career as a transhumanist, I have written with the goal of educating others. If transhumanism is ever going to grow into a mainstream movement, Heaven knows we already have ample scientists and engineers. What the movement really needs now is journalists, politicians, teachers, athletes, activists, artists, and—reluctantly—even lawyers. My writings have aimed to take transhumanism and philosophy of the future out of academia and big tech companies—and into the hands of laypersons.

I believe it is regular people who will lay the foundation to change the world by enlarging the transhumanist movement. We need the future in our daily language, our workplaces, our government, our funerals, our marriages, our hobbies, & maybe even our taxes. We need transhumanism to pervade all activities of our world. That way, we'll always think of the future first, and not get stuck on the past, which is where most of the present seems to be bogged down. The future, as I hope to experience it, is not something that is inevitable. It is something that must be built.

I hope you find strength and inspiration from these essays to help build a futuristic world we can all enjoy!

Zoltan Istvan / December 13, 2020

CHAPTER I: EXPLORING THE FUTURE

1) Post-Earther: Nature Isn't Sacred and We Should Replace It

On a warming planet bearing scars of significant environmental destruction, you'd think one of the 21st Century's most notable emerging social groups—transhumanists—would be concerned. Many are not. Transhumanists first and foremost want to live indefinitely, and they are outraged at the fact their bodies age and are destined to die. They blame their biological nature, and dream of a day when DNA is replaced with silicon and data.

Their enmity of biology goes further than just their bodies. They see Mother Earth as a hostile space where every living creature—be it a tree, insect, mammal, or virus—is out for itself. Everything is part of the food chain, and subject to natural law: consumption by violent murder in the preponderance of cases. Life is vicious. It makes me think of pet dogs and cats, and how it's reported they sometimes start eating their owner after they've died.

Many transhumanists want to change all this. They want to rid their worlds of biology. They favor concrete, steel, and code. Where once biological evolution was necessary to create primates and then modern humans, conscious and directed evolution has replaced it. Planet Earth doesn't need iniquitous natural selection. It needs premeditated moral algorithms conceived by logic that do the most good for the largest number of people. This is something that an AI will probably be better at than humans in less than two decade's time.

Ironically, fighting the makings of utopia is a coup a half century in the making. Starting with the good-intentioned people at Greenpeace in the 1970s but overtaken recently with enviro-socialists who often seem to want to control every aspect of our lives, environmentalism has taken over political and philosophical discourse and direction at the most powerful levels of society. Green believers want to make you think humans are destroying our only home, Planet Earth—and that this terrible action of ours is the most important issue of our time. They have sounded a call to "save the earth" by trying to stomp out capitalism and dramatically downsizing our carbon footprint.

The most important issue of our time is actually the evolution of technology, and environmentalists are mistaken in thinking the Earth is our only or permanent home. Before the century is out, our home for much intelligent life will likely be the microprocessor. We will merge with machines and explore both the virtual and physical universe as sentient robots. That's the obvious destiny of our species and the coming AI age, popularized by past and present thinkers like Stephen Hawking, Ray Kurzweil, and Homo Deus author Yuval Noah Harari. Hundred million dollar companies in California led by billionaires like Elon Musk are already working on technology to directly to connect our brains in real time to the internet. We may soon not need the planet at all, just servers and a power source like solar or fusion.

Even if we somehow don't merge with machines (because scared governments outlaw it, for example), we will still use the microprocessor and its data crunching capabilities to change our genetic make-up so dramatically, that it could not be called: natural. We will enter the Star Wars age where we literally change our DNA and biological appearance to become alien and creature-like—to fit whatever environment we need to fit. If this sounds crazy, just consider the Chinese geneticist who last year changed a girl's genes in utero, creating the first alleged designer baby.

Whatever we become—as a former journalist for the National Geographic Channel who has passionately covered many environmental stories—I want to first make it clear what I think humans are doing to the Earth. I do believe we are destroying the environment. I do think we are overpopulated in many cities. I also believe there is high likelihood humans are helping to cause climate change. And while I do think we should not needlessly destroy the planet (especially wildlife) or live in man-made polluted wastelands, the last thing we need to do is put the brakes on consumption, procreation, and progress.

What we're doing to the planet is not as important as what we are achieving as a species in the nearing of transition to the transhumanist age. We will save and improve far more lives in the future via bioengineering, geoengineering, and coming technology than damaged ecosystems across the planet will harm. Salvation is in science and progress, not sustainability or preserving the Earth. To argue or do otherwise is to be sadistic and act immorally against humanity's well-being.

Besides, the envisioned transhumanist future is not just a place where humans can live without the constant threats and hostility of a biological world, it's an age where sentient beings can finally overcome pain and misery. Beyond shedding our terminal flesh and living indefinitely, a secondary goal of the transhuman movement is to overcome all or the majority of suffering—both for ourselves and other nonhuman animals. This is why some believe transhumanism—even if it's made up of post-earthers—is the most humanitarian movement out there.

The tools transhumanists use—science, technology, and reason—to accomplish its watershed aims rely on thriving economies, free markets, and innovation. These mostly come from competitive countries trying to become powerful and make money—a lot of it. Increased economic output is nearly always responsible for raising the standard of living, something that has been going up a lot in the last 50 years for just about every nation on Earth. But that could change quickly as governments increasingly enforce strict pro-environmental regulation which slows down industry and commerce. When you force companies to operate inefficiently for lofty ideals, it hurts their bottom lines, and that in turn hurts workers and everyday people. It's a well-known fact that when economies slow down, people increasingly lose property, turn to violence, and put having families on hold.

But the media usually doesn't paint environmental policy this way. In fact, the media is responsible for a lot of the misinformation propping up the environmental movement, which is often at odds with transhumanism. A typical news headline reads: Billionaires and Politicians Trying to Protect the Planet. I have to chuckle. Billionaires and politicians usually have power-hungry ambitions. In general, they don't want people to have access to their wealth, power, or pristine environments—because they want it for themselves. That's why they want walls, borders, ownership, and control of it all. How many people without the resources to even afford housing, healthcare, and food will ever take a vacation to protected land, even if the land is public like a national park. How many hundreds of millions of people in inner cities ever go visit "nature"? They don't.

Modern environmentalism is a fabricated deceit of and for the rich and powerful. It's especially prominent in liberal places like New York City and my home town San Francisco. Sadly, environmentalism is often just a terrible tool to wield power over those of lesser means. The amount of minorities that visit US national parks—only 22 percent—compared to whites is totally out of whack.

Despite the imperfections of capitalism, I continue to support it because it remains the best hope for the poor to improve their standard of life—because at least the individually poor can work hard, be smart, and eventually become rich themselves. This rags to riches phenomenon is not something that can happen in socialist or communist environments, where nearly everyone loses (except the corrupt)—and those losses often lead to starvation and eventual civil war.

Enviro-socialists and their green new deals are some of the worst examples of those trying to bring change about to society. These people produce very little—rarely enough to improve society in any meaningful way—and they promise a pristine planet oblivious to the fact the great majority of people will be harmed, not helped, by such economy-killing policies.

However, there is an alternative to this ugly duopoly system we exist in for the masses. Let's harness the capitalists and use our nation's natural resources to end poverty, spread equality, and get humans to the transhuman age where science will make us all healthier and stronger. America has approximately 150 trillion dollars of uninhabited Federal Land not including national parks that we could divide up among its citizens—that's a half million dollars in net worth of resources to every American—all 325 million of them. As a nation, let's sell this federal land or preferably lease it to the capitalists and corporations who can pay us something in return—a permanent universal basic income, for example. Some call this a Federal Land Dividend. Leased properly, our Federal Land could provide over

$1500 a month to every US citizen, giving a household of four $75,000 a year indefinitely.

Naysayers will say the capitalists will forever destroy the land and resources. But over a quarter century, this is unlikely, since all the new capital and innovation from divesting the land will push us far more quickly in the nanotechnology era—an age where we can recreate environments as we please, including those that are destroyed. If you think making plastic with oil is nifty, just wait till we create whole mature forests and jungles in a week's time with coming genetic editing techniques. Also, we'll be able to regrow any animal or plant—including extinct ones—in mass in a laboratory, something that is already on the verge of happening.

But why create the same nature that is so quintessentially cruel, especially as we become transhumans, with perfectly functioning ageless bionic organs and implants in our brains connecting us to the cloud. Let's us create new environments that fit our modern needs. These will be virtual, synthetic, and machines worlds. These new worlds will be far more moral and humanitarian than that of nature. They will be like our homes, cars, and apartments, where everything in it is inanimate or no longer living, and that's why we find sanctuary and comfort in it. If you doubt this, spend a night in the jungle or forest without any comforts or amenities, and see if you survive.

I don't believe in evil, per se, but if there was such a thing, it would be nature—a monster of arbitrary living entities consuming and devouring each other simply to survive. No omnipotent person would ever have the hate in them to create a system where everything wants and needs to sting, eat, and outdo everything else just to live. And yet, that's essentially what the environment is to all living entities. Environmentalists want you to believe nature is sacred and a perfect balance of living things thriving off one another. Nonsense—it's a world war of all life fighting agony and loss—of fight or flight, of death today or death tomorrow for you and your offspring.

It's time to use science and technology to create something better than an environment of biological nature. This begins with admitting the green do-gooder environmentalists are philosophically wrong—and the coming transhumanist age will usher in a world with far less suffering, death, and destruction, even if we have to harm the planet to first get there. Humans must cast off nature, and then they will finally be free of its ubiquitous hostility, misery, and fatalism. Let's rise above the cultural push of environmentalism, because it's antithetical to our future.

****

2) Silicon Valley Wants to Upgrade Pascal's Wager: New Ideas like Quantum Archaeology are Trying to Challenge Religion and Even the Permanence of Death

Since the 17th century, a core part of Western thought has been Pascal's Wager, created by French mathematician and theologian Blaise Pascal. It argues that it makes more sense to believe in God than not, since believing offers a possible way to a happy afterlife, and not believing offers either nothing after death, or maybe even eternal hell. But what happens if humans no longer age or die, something that Silicon Valley and other science hubs around the world are now spending billions of dollars trying to accomplish?

I remember learning about Pascal's Wager in Catholic grade school. As a young Christian, I became fascinated with it and its powerful logical advice. Later, as I grew out of religion and into an agnostic adult, Pascal's Wager formed the antithesis to my own personal philosophy, rendered in my book *The Transhumanist Wager*. The concept of a Transhumanist Wager argues that if it's possible to use science and technology to stop aging and overcome biological death, then that should be the primary goal and direction of one's life. This way, if people are successful in avoiding death, then they don't need to worry about what happens after it—rendering Pascal's Wager obsolete.

Right now, the ways humans are trying to overcome death are somewhat far-fetched and unproven. They involve using drugs and genetic editing to stop and reverse aging; cryonics where frozen dead bodies hope to be brought back to life at a later point when the technology could do it; or even data collecting of media and historical experiences of the deceased—sometimes called Mind

Files—to one day recreate the person as an uploaded cyborg or AI avatar of themselves.

Regardless of the success of life extensionist's current quests to live indefinitely, the complications of Pascal's Wager in the 21st Century era don't end there. Some scientists, philosophers, and even theologians are beginning to argue for a new theoretical philosophy and science called Quantum Archaeology, sometimes referred to as technological resurrection. It's the ability to bring back the dead from any point in history, and it's trying to give a new perspective on spirituality and religion.

Quantum Archaeology has two components: the ability to reverse engineer matter and the ability to 3D bioprint that matter. Based on the trajectory of how fast 3D bioprinting is improving—a tiny human heart was printed out in 2019 by scientists in Israel—it's hard to imagine by year 2100 we wouldn't be able to successfully print out full living human beings. Dr. Tal Dvir at Tel Aviv University's School of Molecular Cell Biology and Biotechnology, who led the team of researchers creating the bioprinted heart, believes by 2029, leading hospitals will already have organ producing printers.

The other component of Quantum Archaeology is more dubious. It relies on the universe being deterministic, which some modern physics like quantum theory has shown unlikely. However, our grasp of the universe and its rules can change as humans become more sophisticated in their research, and plenty of scientists remain on the fence about determinism. Famously, Einstein defended determinism, but more recently astrophysicist Neil deGrasse Tyson said it's possible we live in a simulation, which also suggests a mathematical world of causality. Also, Stanford University theoretical physicist Leonard Susskind wrote a book about how information couldn't be destroyed titled: *The Black Hole War: My Battle with Stephen Hawking to Make the World Safe for Quantum Mechanics*. Naturally, many theologians also believe in some form of determinism because they believe God is omniscient.

Putting the determinism debate aside, let's assume for a moment that humans can at least reverse engineer the past because it has definitely occurred in a specific way. This would be done by massive supercomputers untangling the subatomic occurrences of certain geographical places in the universe—in our case, planet Earth and its surroundings. It sounds insanely large and complex to do this, but already in 2018, America's largest supercomputer was able do 200,000 trillion calculations per second. Quantum Archaeology supporters argue that if the microprocessor can improve for another 100 years following Moore's Law, then the numbers of calculations a supercomputer can do could become greater than the number of atoms that compose them.

The subatomic structural blueprint of a human being in a precise moment in time is big, but not unimaginable. Mike Perry, who has a PhD in Computer Science and has studied the possibilities of technological resurrection, believes that the precise data of all humans who ever lived could fit in a nine square mile databank. The argument goes that if we had the computational power via massive data crunching to reverse engineer and record a human's subatomic structure to discover a point in time—let's say one hour before a person's death—then we could theoretically 3D bioprint those results out and that person's human life would be restored, exactly as it once was. The mind, body, and even memories would be precisely the same person, down to very molecules and atoms.

As a result of these speculative ideas, there are organizations, such as the Society for Universal Immortalism and the Turing Church—that embrace bringing back every dead person who has ever lived—all 100 billion of us since we became homo sapiens. There are also Christian Transhumanists who believe in creating Quantum Archaeology so that when Jesus makes his Second Coming to Earth, this tech might be used to facilitate saving believers. Reincarnationists, such as some Hindus, are another group that are open to the idea of technological resurrection.

I'm not a believer in Quantum Archaeology. Not yet at least. Bringing back the dead is not only wildly complicated but reeks of ethical controversy. For example, it would create far worse overpopulation issues on Earth. There would be 3D bioprinted spouses meeting their loved ones who may have remarried and had new families. There would be potentiality for misuse of the technology; dictators might bioprint out clone armies of themselves and family members. And who would get resurrected? What if some people did who didn't want to be brought back to life? What would they do? And in case Quantum Archaeology isn't bizarre enough: Should we begin to put "Do not resurrect me" or "Please resurrect me" in our wills?

On the other hand, who doesn't want to bring back a loved one they lost. My father died recently from cardiac failure—he had already had four heart attacks in the two decades before—and I would love to bring him back to life. Like me, he was agnostic and didn't believe in an afterlife, but I'm certain he wanted to live longer, if his health could be improved. Naturally, that's the promising thing about the possibility of such exceptional 3D bioprinting technology. Theoretically, one's brain could be printed out identically to what it was, but one's other organs could be improved to be disease-free and rejuvenated. And it's likely by year 2100, other new life extension and anti-aging therapies would also exist to help make people younger again.

Regardless of the ethics and whether the science can even one day be worked out for Quantum Archaeology, the philosophical dilemma it presents to Pascal's Wager is glaring. If humans really could eradicate the essence of death as humans know it—including even the ability to ever permanently die—Pascal's Wager becomes unworkable. Frankly, so does my Transhumanist Wager. After all, why should I dedicate my life and energy to living indefinitely through science when by next century technology could bring me back exactly as I was—or even as an improved version of myself?

Outside of philosophical discourse, billions of dollars are pouring into the anti-aging and technology fields—much of it from Silicon Valley. Everyone from entrepreneurs like Mark Zuckerburg to nonprofits like XPRIZE to giants like Google are spending money on ways to try to end all disease and overcome death. Bank of America recently reported that they expect the extreme longevity field to be worth over $600 billion dollars by 2025.

Technology research spending for computers, microprocessors, and information technology is even bigger: $3.7 trillion dollars is estimated to be spent worldwide in 2019. This amount includes research into quantum computing, which is hoped to eventually make computers hundreds—maybe thousands—of times faster over the next 50 years.

Despite the advancements of the 21st Century, the science to overcome biological death is not very close to being ready. Over 100,000 people still die a day, and in some countries like America, life expectancy has actually started going slightly backward. However, like other black swans of innovation in history—such as the internet, combustion engine, and penicillin—we shouldn't rule out that new inventions may make humans live dramatically longer and maybe even as long as they like. As our species reaches for the heavens with its growing scientific armory, Pascal's Wager is going to be challenged. It just might need an upgrade soon.

****

3) Programming Hate into AI Will be Controversial, but Probably Necessary

In the last few years, the topic of artificial intelligence (AI) has been thrust into the mainstream. No longer just the domain of sci-fi fans, nerds or Google engineers, I hear people discussing AI at parties, coffee shops and even at the dinner table: My five-year-old daughter brought it up the other night over taco lasagna. When I asked her if anything interesting had happened in school, she replied that her teacher discussed smart robots.

The exploration of intelligence — be it human or artificial — is ultimately the domain of epistemology, the study of knowledge. Since the first musings of creating AI back in antiquity, epistemology seems to have led the debate on how to do it. The question I hear most in this field from the public is: How can humans develop another intelligent consciousness if we can't even understand our own?

It's a prudent question. The human brain, despite being only about three pounds in weight, is the least understood organ in the body. And with a billion neurons — with 100 trillion connections — it's safe to say it's going to be a long time before we end up figuring out the brain.

Generally, scientists believe human consciousness is a compilation of many chemicals in the brain forced though a prism that produces cognitive awareness designed to insist an entity is aware of not only itself but also the outside world.

Some people argue that the quintessential key to consciousness is awareness. French philosopher and mathematician René Descartes may have made the initial step by saying: I think, therefore I am. But thinking does not adequately define consciousness. Justifying thinking is much closer to the meaning that's adequate. It really should be: I believe I'm conscious, therefore I am.

But even awareness doesn't ring right with me when searching for a grand theory of consciousness. We can teach a robot all day to insist it is aware, but we can't teach it to prove it's not a brain in a vat — something people still can't do either.

Christof Koch, chief neuroscientist at the Allen Institute for Brain Science, offers a more unique and holistic version of consciousness. He thinks consciousness can happen in any complex processing system, including animals, worms and possibly even the Internet.

In an interview, when asked what consciousness is, Koch replied, "There's a theory, called Integrated Information Theory, developed by Giulio Tononi at the University of Wisconsin, that assigns to any one brain, or any complex system, a number — denoted by the Greek symbol Φ — that tells you how integrated a system is, how much more the system is than the union of its parts. Φ gives you an information-theoretical measure of consciousness. Any system with integrated information different from zero has consciousness. Any integration feels like something."

If Koch and Tononi are correct, then it would be a mistake to ever think one conscious could equal another. It would be apples and oranges. Just like no snowflake or planet is the same as another, we must be on our guard against using anthropomorphic prejudice when thinking about consciousness.

In this way, the first autonomous super-intelligence we create via machines may think and behave dramatically different than us — so much so that it may not ever relate to us, or vice versa. In fact, every AI we ever create in the future may leave us in very short order for distant parts of the digital universe — an ego-thumping concept made visual in the brilliant movie *Her*. Of course, an AI might just terminate itself, too, upon realizing it's alive and surrounded by curious humans peering at it.

Whatever happens, in the same way there is the anthropological concept cultural relativism, we must be ready for "consciousness relativism" — the idea that one consciousness may be totally different than another, despite the hope that math, logic and coding will be obvious communication tools.

This makes even more sense when you consider how small-minded humans and their consciousness might actually be. After all, nearly all our perception comes from our five senses, which is how our brain makes sense of the world. And every one of our senses is quite poor in terms of possible ability. The eye, for example, only sees about 1 percent of the universe's light spectrum.

For this reason, I'm reluctant to insist on consciousness being one thing or the other, and do lean toward believing Koch and Tononi are correct by saying variations of consciousness can be seen in many forms across the spectrum of existence.

This also reinforces why I'm reluctant to believe that AI will fundamentally be like us. I surmise it may learn to replicate our behavior — perhaps even perfectly — but it will always be something different. Replication is no different than the behavior of a wind-up doll. Most humans hope for much more of themselves and their consciousness. And, of course, most AI engineers want much more for the machines they hope to give a conscious rise to.

Despite that, we will still try to create AI with our own values and ways of thinking, including imbuing it with traits we possess. If I had to pinpoint one behavioral trait of consciousness that humans all have and should also be instilled in AI, it would be empathy. It's empathy that will form the type of AI consciousness the world wants and needs — and one that people also can understand and accept.

On the other hand, if a created consciousness can empathize, then it must also be able to like or dislike — and even to love or hate something.

Therein lies the conundrum. In order for a consciousness to make judgments on value, both liking and disliking (love and hate) functions must be part of the system. No one minds thinking about AI's that can love — but super-intelligent machines that can hate? Or feel sad? Or feel guilt? That's much more controversial — especially in the drone age where machines control autonomous weaponry. And yet, anything less than that coding in empathy to an intelligence just creates a follower machine — a wind-up doll consciousness.

Kevin LaGrandeur, a professor at the New York Institute of Technology, recently wrote, "If a machine could truly be made to 'feel' guilt in its varying degrees, then would we have problems of machine suffering and machine 'suicide'"? If we develop a truly strong artificial intelligence, we might — and then we would face the moral problem of creating a suffering being.

It's a pickle for sure. I don't envy the programmers who are endeavoring to bring a super intelligence into our world, knowing that their creations may also consciously hate things — including its creators. Such programming may just lead to a world where robots and machine intelligences experience the same modern-day problems — angst, bigotry, depression, loneliness and rage — afflicting humanity.

****

4) Future Transhumanist Tech May Soon Change the Definition of Disability

Radical technologies around the world may soon overhaul the field of disability and immobility, which affects in some way more than a billion people around the world.

MIT bionics designer Hugh Herr, who lost both his legs in a mountain climbing accident, recently said in a TED Talk on disability, "A person can never be broken. Our built environment, our technologies, are broken and disabled. We the people need not accept our limitation, but can transcend disability through technological innovation."

His words are coming true. Around the world, the deaf hear via cochlear implants, paraplegics walk with exoskeletons and the once limbless have functioning limbs. For example, some amputees have mind-controlled robotic arms that can grab a glass of water with amazing precision. In 15 or 20 years, that bionic arm could very well be better than the natural arm, and people may even electively remove their biological arms in favor of robotic ones. After all, who doesn't want to be able to do a hundred pull ups in a row or lift the front end of a car up to quickly change a flat tire?

The same radical improvements will happen with eyesight. Already, blind people can see some things better than the natural eye with robotic technology made possible by the Argus II. In 15 years time, expect bionic eyes to be electively installed in our eye sockets, as they will be superior to human eyes. And wheelchairs? They are likely going the way of the dinosaurs. Expect some wheelchair companies to go bankrupt if they don't diversify over the next decade as exoskeleton technology becomes commonplace.

Soon, I think many people will have exoskeleton suits — disabled or not. In fact, some people will probably have different models. Some suits will be for sports, some for wearing exclusively on the battlefields (such as the Ironman suit) and some will be just for having crazy sex in positions most people never thought possible.

All this is great news for the hundreds of millions of disabled people around the world, many who are suffering and are mentally depressed as a result of their handicaps. Transhumanist technology is revolutionizing the way we deal with physical disability. And I couldn't be happier about that fact. I hope every handicapped person in a wheelchair gets out of it and into an exoskeleton suit. And I hope in two decade's time, improved stem cell technology will get their spines and other nonfunctioning body parts back to ideal health and strength.

The medical field will probably have to come up with new terms to define this post-disability age.

Writer B. J. Murphy recently emailed me, "In the future where mostly everyone is able bodied, how might we define 'disability?' Surely, as the disabled become augmented and/or enhanced, our current definition of what makes someone disabled — not able bodied — will be done away with altogether. Though, how do we then consider those who are, albeit able-bodied, not augmented or enhanced? Do we change the definition of 'disability' as something which someone simply is when not modified with advanced technologies?"

The questions are certainly interesting and important. Could we one day be a species that sees itself as born "not fully able-bodied," or even technically disabled before modern technology has had a chance to radically transform us? Exoskeleton suits and bionic boots will probably allow us to run as fast as cheetahs in the near future, and bionic eyes will allow us to see far more of the existing light spectrum than human eyes can, so such ideas are possible.

Continuing with this line of thinking, people who don't use smartphones, own and drive a car or participate in social media are sometimes seen at a definite disadvantage in the 21st century — one that some might call a social handicap and that some employers would find unhireable.

On the flip side, though, some people will probably be turned off by too much augmentation and technology in their lives, even when it helps their disabilities. And those people may refuse treatment or help. That is, of course, their right to do so, and it will be important to respect their feelings and prerogatives. But many others will welcome radical technology to change their lives and physical states.

"It's not well appreciated," said, Herr in his TED Talk, "but over half of the world's population suffers from some form of cognitive, emotional, sensory or motor condition, and because of poor technology, too often, conditions result in disability and a poorer quality of life. Basic levels of physiologically function should be a part of our human rights. Every person should have the right to live without disability if they so choose."

Regardless of how people accept such enhancing technology, I'm betting in the future, the definition of disability will change dramatically. It will be based on what kind of improvements people accept and what kinds they do not accept. The "disability" definition will probably change as much as the definition of death has been changing over the last decade. With suspended animation now being practiced in an American hospital in Pittsburgh, doctors can keep people completely dead for a few hours, then revive them back to life. In five years, that "few" hours are likely to become 24 hours.

And in 10 years, that brain-dead suspended animation state could become weeks or even months (which futurists hope will be useful for space travel). Like the changing specter of death, the challenge of physical disability upon the human race will also change. If the species allows these changes, we may overcome physical disability completely.

The medical field will probably have to come up with new terms to define this post-disability age that transhumanists aim to usher in. Those new terms will be welcomed words for many of the hundreds of millions of disabled people around the world eagerly waiting to regain their full physical possibilities.

****

5) What If One Country Achieves the Singularity First?

The concept of a technological singularity is tough to wrap your mind around. Even experts have differing definitions. Vernor Vinge, responsible for spreading the idea in the 1990s, believes it's a moment when growing superintelligence renders our human models of understanding obsolete. Google's Ray Kurzweil says it's "a future period during which the pace of technological change will be so rapid, its impact so deep, that human life will be irreversibly transformed." *Gizmodo* Editor-in-Chief Annalee Newitz explained, "A good way to understand the singularity is to imagine explaining the internet to somebody living in the year 1200." Even Christian theologians have chimed in, sometimes referring to it as "the rapture of the nerds."

My own definition of the singularity is: the point where a fully functioning human mind radically and exponentially increases its intelligence and possibilities via physically merging with technology.

All these definitions share one basic premise—that technology will speed up the acceleration of intelligence to a point when biological human understanding simply isn't enough to comprehend what's happening anymore.

That also makes a technological singularity something quasi-spiritual, since anything beyond understanding evokes mystery. It's worth noting that even most naysayers and luddites who disdain the singularity concept don't doubt that the human race is heading towards it.

In March 2015, I published a *Vice Motherboard* article titled *A Global Arms Race to Create a Superintelligent AI is Looming*. The article argued a concept I call the AI Imperative, which says that nations should do all they can to develop artificial intelligence, because whichever country produces an AI first will likely end up ruling the world indefinitely, since that AI will be able to control all other technologies and their development on the planet.

The article generated many thoughtful comments on Reddit Futurology, LessWrong, and elsewhere. I tend not to comment on my own articles in an effort to stay out of the way, but I do always carefully read comment sections. One thing the message boards on this story made me think about was the possibility of a "nationalistic" singularity—what might also be called an exclusive or private singularity.

If you're a technophile like me, you probably believe the key to reaching the singularity is two-fold: the creation of a superintelligence, and the ability to merge humans with that intelligence. Without both, it's probably impossible for people to reach it. With both, it's probably inevitable.

Currently, the technology to merge the human brain with a machine is already underway. In fact, hundreds of thousands of people around the world already have brain implants of some sort, and a few years ago telepathy was performed between researchers in different countries. Thoughts were passed from one mind to another using a machine interface, without speaking a word.

Fast forward 25 years in the future, and some experts believe we might already be able to upload our entire consciousness into a machine. I tend to agree, and I even think it could occur sooner, such as in 15 to 20 years' time.

Here's the crux: If an AI exclusively belonged to one nation (which is likely to happen), and the technology of merging human brains and machines grows sufficiently (which is also likely to happen), then you could possibly end up with one nation controlling the pathways into the singularity.

As insane as this sounds, it's possible that the controlling nation could start offering its citizens the opportunity to be uploaded fully into machines, in preparation to enter the singularity. Whether there would then be two distinct entities—one biological and one uploaded—for every human who chooses to do this is a natural question, and it's only one that could be decided at the time, probably by governments and law. Furthermore, once uploaded, would your digital self be able to interact with your biological self? Would one self be able to help the other? Or would laws force an either-or situation, where uploaded people's biological selves must remain in cryogenically frozen states or even be eliminated altogether?

No matter how you look at this, it's bizarre futurist stuff. And it presents a broad array of challenging ethical issues, since some technologists see the singularity as something akin to a totally new reality or even a so-called digital heaven. And to have one nation or government controlling it, or even attempting to limit it exclusively to its populace, seems potentially morally dubious.

For example, what if America created the AI first, then used its superintelligence to pursue a singularity exclusively for Americans?

(Historically, this wouldn't be that far off from what many Abrahamic world religions advocate for, such as Christianity or Islam. In both religions, only certain types of people get to go to heaven. Those left behind get tortured for eternity. This concept of exclusivity is the single largest reason I turned nonreligious when I was 18 years-old.)

Worse, what if a government chose only to allow the super wealthy to pursue its doorway to the singularity—to plug directly into its superintelligent AI? Or what if the government only gave access to high-ranked party officials? For example, how would Russia's Vladimir Putin deal with this type of power? And it is a tremendous power. After all, you'd be connected to a superintelligence and would likely be able to rewrite all the nuclear arms codes in the world, stop dams and power plants from operating, and create a virus to shut down Wi-Fi worldwide, if you wanted.

Of course, given the option, many people would probably choose not to undergo the singularity at all. I suspect many would choose to remain as they are on Earth. However, some of those people might be keen on acquiring the technology of getting to the singularity. They might want to sell that tech, and offer paid one-way trips for people who want to have a singularity. For that matter, individuals or corporations might try to patent it. What you'd be selling is the path to vast amounts of power and immortality.

Such moral leanings and concepts that someone or group could control, patent, or steal the singularity ultimately lead us to another imperative: the Singularity Disparity.

The first person or group to experience the singularity will protect and preserve the power and intelligence they've acquired in the singularity process—which ultimately means they will do whatever is necessary to lessen the power and intelligence accumulation of the singularity experience for others. That way the original Singularitarians can guarantee their power and existence indefinitely.

In my philosophical novel *The Transhumanist Wager*, this type of thinking belongs to the Omnipotender, someone who is actively seeking and contending for as much power as possible, and bases their actions on such endeavors.

I'm not trying to argue any of this is good or bad, moral or immoral. I'm just explaining how this phenomena of the singularity likely could unfold. Assuming I'm correct, and technology continues to grow rapidly, the person who will become the leading omnipotender on Earth is already born.

Of course, religions will appreciate that fact, because such a person will fulfill elements of either the Antichrist or the Second Coming of a Jesus, which is important to both the apocalyptic beliefs in Christianity and Islam. At least the "End Times" are really here, faith-touters will be able to finally say.

The good news, though, is that maybe a singularity is not an exclusive event. Maybe there can be many singularities.

A singularity is likely to be mostly a consciousness phenomenon. We will be nearly all digital and interconnected with machines, but we will still able to recognize ourselves, values, memories, and our purposes—otherwise I don't think we'd go through with it. On the cusp of the singularity, our intelligence will begin to grow tremendously. I expect the software of our minds will be able to be rewritten and upgraded almost instantaneously in real time. I also think the hardware we exist through—whatever form of computing it'll be—will also be able to be reshaped and remade in real time. We'll learn how to reassemble processors and their particles in the moment, on-demand, probably with the same agility and speed we have when thinking about something, such as figuring out a math problem. We'll understand the rules and think about what we want, and the best answer, strategy, and path will occur. We'll get exceedingly efficient at such things, too. And at some point, we won't see a difference between matter, energy, judgment, and ourselves.

What's important here is the likely fact that we won't care much about what's left on Earth. In just days or even hours, the singularity will probably render us into some form of energy that can organize and advance itself superintelligently, perhaps into a trillion minds on a million Earths.

If the singularity occurs like this, then, on the surface, there's little ethically wrong with a national or private singularity, because other nations or groups could implement their own in time. However, the larger issue is: How would people on Earth protect themselves from someone or some group in the singularity who decides the Earth and its inhabitants aren't worth keeping around, or worse, wants to enslave everyone on Earth? There's no easy answer to this, but the question itself makes me frown upon the singularity idea, in exactly the same way I frown upon an omnipotent God and heaven. I don't like any other single entity or group having that much possible power over another.

****

6) Why I'm Debating an Anarcho-Primitivist Philosopher About the Future

Imagine if you woke up one morning and nearly all technological advancement developed over the last few millennia was gone from the Earth. For starters, your bed in your cave would likely be an uncomfortable pile of hay and dried animal skins. You'd smell pretty bad too, since there wouldn't be much incentive to bathe without hot running water, shampoo, or clean towels. And if you were elderly at all, it's likely some place on your body might hurt, such as a tooth with a cavity—and because you don't have aspirin and antibiotics to get through an infection, you might die from it.

Does all this sound crazy? Anarcho-primitivists—"people who advocate for a return to non-'civilized' ways of life through deindustrialization," according to Wikipedia—want you to embrace it. They say technology and progress are increasingly a downer, which only serves to alienate us from… well… just about everything in the post-Caveman era.

If you think I'm going to bash anarcho-primitivists, you're wrong. As a transhumanist, I totally agree with their overarching point that technology and progress is fundamentally changing the human being for the worse. If you're a human and want to remain a human, then technology and civilization has gotten way out of control and is not only going to drive you nuts, but also destroy your way of life. I see it happening all the time to people—especially older conservative people—and it's no fun for them. Human brains and bodies were not made for so much radical technology, for the concrete mega-cities many of us live in, or for the busy labor-intensive social schemes we participate in. However, transhuman beings—those who like to drive cars, fly on jet airplanes, surf the internet, go scuba diving, climb Mt. Everest using oxygen tanks, live in skyscrapers, and use a flushing toilet—the world is getting better for them. Much better.

Transhuman means beyond human. And frankly, transhumanists are people who are just not interested in being humans anymore. We are not fans of our fragile flesh, of our mortal bodies, or of our flawed organs, such as our eyes which can see a mere 1 percent of the light spectrum of the universe. We are not interested in dying from cavities, in sleeping in caves, in throwing spears at wildlife, and in worrying about the survival trifles that have affected the species for hundreds of thousands of years. We aim to leave humanity behind and embrace a tech and science-dominated future, full of cyborg body parts, digital environments, indefinite lifespans, and new social philosophies.

The fundamental conundrum with anarcho-primitivism versus transhumanism is that people tend to believe we are still human—that we are still mammals, and primates at that. Such a simplified description of us may be helpful when teaching a third-grader in elementary school, but it's quickly loses its relevance when discussing a person wearing an exoskeleton suit, possessing a tracking RFID chip, taking Viagra, and wearing Google Glass. Are we human? Not so much anymore. We started crossing that bridge a long time ago when we received our first vaccine, or made a phone call, or watched an astronaut walk on the moon.

This weekend at Stanford University I have the honor of publicly debating the world's leading anarcho-primitivist philosopher John Zerzan. The idea of the debate began with *The Telegraph* journalist Jamie Bartlett's new nonfiction book *The Dark Net*. The closing chapter of the book is called Zoltan vs. Zerzan, and Barlett juxtaposes our two rival philosophies in it.

Just before the book was published, Bartlett suggested that a public debate might be a good way to put a finger on the pulse of progress in the world from two polar opposite views. Zerzan and I agreed, and the Stanford Transhumansit Association—one of the largest student futurist associations in America—kindly signed up to host the debate. The event, which is in the Geology Corner Auditorium (320-105) this Saturday night, November 15th, at 7 PM, is free and open to the public.

Naturally, as civilization evolves into greater technological complexity, the conflict between anarcho-primitivism and transhumanism grows every year, too.

Recently, Hollywood released the blockbuster film *Transcendence*, starring Johnny Depp. Part of the movie highlights an anti-technology organization against transhumanist scientists.

Transhumanists wish it were only fiction, but the conflict occurs in real life, as well. Recently, Italian and Mexican anti-technology groups mailed bombs to scientists and technologists in Switzerland and Mexico. Thankfully, John Zerzan appears not to be connected to these groups, but his past history as a public confidant of the Unabomber and someone possibly involved with helping to orchestrate the Seattle Riots has made me cautious. If I look a little uncomfortable during my debate, it's because I might be wearing a bulletproof vest, which unfortunately will occasionally become a part of my new dress code as the 2016 US presidential candidate of the recently formed Transhumanist Party.

Up until recently, I believed it was religious fundamentalists who were the main obstacles to getting society to embrace transhumanism and the coming radical technological change that is sweeping over our species. However, over the last two years I've come to realize that anarcho-primitivists are also formidable foes to the transhuman future. I still worry more about religious conservatives wanting to stop progress (as George W. Bush did with stem cell research during his time in office), but eco-friendly activists & "Occupy" supporters also have power and large numbers around the world. And they must be considered as being potentially disruptive to transhumanism and its growth.

The big issue with anarcho-primitivists is that they have ties to various other fringe groups, which collectively might put their supporters in the millions. The Earth Liberation Front (ELF), various anarchist organizations, and the Occupy movements all hold various philosophies that can to some extent be linked with anarcho-primitivism. The danger here is that the greater coalition of anarcho-primitivists end up convincing the public that in order to save the Earth, tackle overpopulation, and address environmental issues—certainly some of the most hot-button topics of the year—we must take big steps backward with progress and civilization. Environmentalists taking up the banner of anti-technology would not be good recipe for a successful, forward-thinking society.

The support that anarcho-primitivists bring to green issues is the most important point they have in their arsenal of influence. Moderate anarcho-primitivists advocate for a major slowdown of technological development and capitalistic systems that care more for profits than they do for people or the planet. On the surface, this seems like a sound point. However, the flipside is that by slowing down technology and commerce, you would inevitably also be shortening and harming the lives of people, including some of the most vulnerable in society. For example, the BBC reports that girls and women were hit hardest by the recent global recession, causing higher rates of abuse, starvation, and infant mortality.

A recent report from Henry Blodget at *Business Insider* points out that today people all around the world are living better than they ever have. Poverty rates are lower. Lifespans are longer. Fewer people on average are starving to death. More opportunity for jobs exist. Violent war deaths are declining. Twenty-first century science, technology, and globalization are responsible for these improvements in the world. Transhumanists strongly condemn those who advocate halting the very things that have helped give people more food, longer lives, and better health.

Furthermore, as much as most people care about the environment and the Earth, people care more about how long and well they live, especially when it concerns their loves ones too. To stop technological innovation today would ultimately harm the lives of all people in all countries. Transhumanists advocate for green-friendly use of technology and the creation of technological fixes to help the world's growing environmental issues. These are the actions that will help the world find the proper balance with Earth and its environment.

In 1995, I journeyed far into the mountains of Espiritu Santo in Vanuatu and became the first foreigner ever to visit the tribe of Mareki. It was a truly traditional village living as some of the tribes had done for centuries before. In 2002, I returned to Vanuatu to film that same tribe for the National Geographic Channel. While both my stays in the village were amazing and magical, much of my experience was tempered with sadness. With no medicine or technology, villagers often died young from disease, malnutrition, and infection. People in the village told me little more than half of the children born survive to adulthood. Communities like that are in a constant state of mourning for their lost loves ones and a constant state of survival for their own lives.

I worry that anarcho-primitivists romanticize lifestyles like that of the Mareki tribe. While there are beautiful moments to be had in all ways of life, a journey into anti-civilization ways would quickly morph into a desperate and exhausting quest of survival. Our species has come way too far to abandon progress and go backwards. The future is forward, and it's a transhuman future, filled with health, well-being, and techno-optimism.

****

7) Do We Have Free Will Because God Killed Itself?

Some people believe humans with our three-pound brains are the most advanced life form ever to exist; I am not one of them. To insist we are alone in the universe, or that we are the galaxy's crowing civilization, reeks of ego—and reminds me of those who insisted the Earth was flat.

The universe is 13.8 billion years old according to experts. A lot can happen in so much time, such as the rise (and fall) of superintelligences amongst the approximately two billion life-friendly planets that exist in our galaxy.

It is likely that these highly advanced intelligences long ago reached what we call the singularity: a moment in time when technological acceleration—most likely through the creation of artificial superintelligences—becomes incredibly rapid.

This presents a thorny issue to humans because of what I call the Singularity Disparity—the idea that whoever reaches the singularity first will make sure no one else can achieve a similar amount of power.

If we are not alone in the universe and also not the most intelligent life forms, then it's unlikely our species can evolve beyond a certain point, since other more advanced life forms won't allow it.

So where does that leave us, a species about 20 to 40 years away from building superintelligences that will help us reach the singularity? The answer is not pretty. In fact, if I had to guess—based on some of the recent discoveries in string theory—we're likely already existing in some type of simulation created by an ancient superintelligence, one where we're observed, regulated, and possibly even manipulated at times.

Worse, other superintelligences likely structured the intelligences controlling us before them, and so on.

I'm not going to argue the merits of whether we live in a simulated hologram universe or not, all of which have been covered by philosophers through the ages, from Aristotle to Oxford's Nick Bostrom to John Searle and his Chinese room. Suffice to say, there's enough scientific and philosophical evidence for me to slightly tilt in favor of it all. For me, however, the more interesting question is why would we live in a simulation? Given the Singularity Disparity, why would some superintelligence or group of superintelligences do this to us?

There are various explanations. The main ones are:

1) We are experiments and playthings for those superintelligences using us to further understand themselves or support some causes of theirs, including dealing with boredom.

2) We are literally already intrinsic parts of those superintelligences and exist simply as their thoughts, energies, or structure (the Gaia people love this idea).

3) We are accidents in the universe and our existence is totally arbitrary.

The deity-averse existentialist in me likes #3 best, but I'm still not satisfied with any of the answers, mainly because none of them address what happened to the very first superintelligence, an entity who may have set all the universe's rules up.

Luckily, there is a fourth, more controversial take that I do think is worth exploring. The foundation of the universe, including all the simulations, probabilities, and possibilities of existence are the result of the first and most powerful superintelligence killing itself.

In short, an entity literally on the verge of becoming God knowingly and willingly died by suicide.

The problem with being God—a truly omnipotent being—is that of free will. As a recent comedy skit called Future Christ on *The Daily Show* with Jon Stewart—a skit which partially resulted from my original atheistic story—pointed out: "If God wants to quit smoking, can he hide cigarettes from himself?"

Being all-powerful is a very strange, ironic dead end. The only thing omnipotence can truly equal is a total mechanistic void. Achieving omnipotence is literally the act of suicide, in terms of forever self-eliminating one's consciousness. This is because a conscious intelligence, as reason dictates, is based on ability to discern values—values, for example, to know whether as an all-powerful being, one can create something so heavy that one can't lift it. Values require choice. But omnipotence means that all choices have already been made, and nothing can ever change, because all variables are already accounted for and no randomness or anomalies exist.

It's quite possible, a long time ago, that the first superintelligent Singularitarian decided to up its game and attempted to become omnipotent. But if it succeeded—and it may have—then it would have become an entity without a singular intelligent consciousness, because intelligence requires choice. For all practical matters, it would cease to exist in a personal interactive way that any intelligence could relate to.

But before this first Singularitarian did that, it would've left us with its rules—physical laws of the universe that contain our potential power and intelligence. It would've also left us with the code of the Singularity Disparity, where the singularity we achieve will never equal other singularities or be the most powerful.

If this is all the case, this leaves the human race in a precarious position. Here we are, in a universe where many singularities have almost certainly taken place, but reaching anything beyond a certain point becomes impossible due to limits of pre-existing natural laws. Adding to the mix are other superintelligences that don't want us to dominate or overpower them, either, just as we don't want any other entities on Earth to dominate or overpower us. Hierarchies and power plays exist everywhere—they are the fabric of the universe.

As an atheist (or even a possible theistcideist—one who believes God or a supreme being once existed but no longer does because it terminated itself), I would commend this leading superintelligence for destroying its conscious self. By doing so, and establishing that nothing else could ever become as powerful as itself, it would've forever sown choice into the universe, since no one can ever reach a perfect position of choice-less omnipotence, and the death of its consciousness would mean it couldn't ever change what it had done. This superintelligence's final acts have assured all other advanced life forms the possibility of free will and the ability to try to become more than we are.

****

8) Will Capitalism Survive the Coming Robot Revolution?

Economic experts are trying to figure out a question that just two decades ago seemed ridiculous: If 90 percent of human jobs are replaced by robots in the next 40 years — something now considered plausible — is capitalism still the ideal economic system to champion? No one is certain about the answer, but the question is making everyone nervous — and forcing people to dig deep inside themselves to discover the kind of future they want.

After America beat Russia in the Cold War, most of the world generally considered capitalism to be the hands-down best system on which to base economies and democracies. For decades, few doubted capitalism's merit, which was made stronger by thriving globalization and a skyrocketing world net worth. In 1989 — when the Berlin Wall fell — the world had only 198 billionaires. Now, according to Forbes, there are 1,826 of them in 2016.

Despite growing riches, when banks collapsed in 2007 during the Great Recession, the world stepped back and wondered aloud if a more nuanced approach to economic progress was needed. These doubts of 21st century capitalism helped set the stage for an economic paradigm shift just starting to appear — economists observing jobs not just disappearing to other countries, but disappearing off the face of the Earth. The culprit: robots and software.

At first, the warnings of this weren't very loud. After all, economies and companies thrive because of modernization, which includes upgrading with new tech to make and save money. But in the last year, a growing chorus of people are beginning to see a tipping point, maybe a decade in the future, where tens of millions of jobs may be lost in as short as a five-year period — which would be many more times the jobs lost during the Great Recession.

Already today, there are countries trying out driverless trucks to deliver goods. Truck driving is one of the most prevalent jobs in America, with about 3.5 million drivers. What will we do in five years if they are replaced with vehicles that don't need human intervention to get on and off a highway to deliver goods?

Of course, they are just one occupation amongst many dozens — like waiters, bank tellers and even librarians — that might no longer require humans in the very near future.

Capitalism says this is the nature of the competitive economy. However, those jobs that are replaced will never be regained, and truck drivers and waiters will not easily find other jobs. Many will likely need to be provided for by the state, otherwise grown men and women will surely pick up Molotov cocktails and show the world a thing or two about worker revolutions.

The only difference between this and other historical revolutions is they won't be alone. This time it's not a problem of the rich versus the poor. In 20 years, everyone's job will be at stake, even that of my wife, who trained 19 years in college to become a practicing Ob/Gyn — and still today has $100,000+ in school debt. But machines will deliver babies and remove cervical cancer better than people. And software will do taxes more efficiently than accountants. And articles will be crafted better by news aggregating software than living, breathing journalists.

Everyone, including even the U.S. president, is at risk of being outperformed by a machine — and eventually being jobless and without income.

So, now that we know we're all going to lose our jobs, what system can make it so humans will still be happy and live better without employment? Clearly, it's not capitalism.

Whichever system we choose will have to incorporate an improving standard of life for people and society. For this reason, I tend to support a Universal Basic Income as one way to desire robots to take our jobs but not leave the world poor. However, that doesn't really say what will happen to economies after the robot revolution is really underway.

Some people have said a fully automated luxury communism will prevail once robots take all the jobs — an economic system that favors technology pampering humans all day long. Communism is a historically loaded word that few people like (including myself, a longtime entrepreneur).

Additionally, it insinuates being chained to community and social service, something I think our individualistic-minded world may scoff at. The 21st century has made people feel more entitled than ever, and, frankly, with so much amazing innovation humans have come up with, we deserve it. We deserve to be pampered by technology. We deserve to never again work a day in our lives if we don't want to. We deserve not to be bothered by government or society if we're not bothering others. And we deserve to pursue lofty dreams instead struggling to earn a handful of dollars.

In fact, I doubt money will even survive this century. If anything, in the future, only knowledge will have tradable value — the knowledge to create better machines, software and experiences from technology. Around this time — surely before 2075 — the singularity will be possible, a point where people connect themselves to artificial intelligence and essentially disappear into a sea of growing and organizing information. Then it's anyone's guess what happens to the world.

However, back to reality here in 2016: Whatever economic system does prevail in the next 25 years, it won't be like anything we thought of before. Karl Marx and Adam Smith simply did not account for what indefinite robot labor would mean to a new world increasingly reliant on microprocessors and 1s and 0s for its every step forward.

Whatever happens, it's probably best to keep an open mind about the future and new economic models. Many of us are running on a financial treadmill right now, trying to get ahead and realize the American Dream of riches and the good life. But in the future, the American Dream may be more about discovery of our newly acquired transhuman possibilities and enjoying the technology that has made our modern lives so simple and easy. I think I can get used to that.

****

9) The Culture of Transhumanism is About Self-improvement

Over the last few years, I've received various reactions from the public about my articles on transhumanism. Those reactions have ranged all across the board—from spewing hatred to mocking skepticism to genuine interest. The thing with transhumanism—the core of its message—is whatever it espouses, it's new thinking. Whether its brain implants, bionic limbs, designer babies, robotic hearts, exoskeleton suits, artificial intelligence, or gene therapies that aim to eliminate biological death, it's decidedly uncharted territory for the human species.

For better or worse, the society we live in treats virtually anything uncharted and socially new as contentious. We are cultured to be afraid and hesitant. In fact, being cultured (and the participation in such culture) is really just the acceptance and practice of standards, ideas, and ideals pre-established by others. Generally, the more a certain culture entrenches itself within its ways, the less chance it has to be new and dynamic. Most cultures—whether religious, national, or social—aim to uphold and solidify their values and idiosyncrasies.

Transhumanism doesn't do that. It's not beholden to anything idiosyncratic. It wants to improve the human body with science and technology—which is to say it wants to help people evolve. That's a strange cultural and philosophical position for a movement. And yet, evolution is exactly what transhumanism aims to usher in—except transhumanists want to do it far more quickly than by glacial-paced natural selection.

Of course, some people will ask: How far will transhumanism go to evolve the species? Will it support merging humans with machines? Will it eliminate biological death forever? Is it atheistic or spiritual? Will it be responsible for changing the human race into something foreign?

I certainly don't know all these answers. But I do know the core of transhumanist culture is it's perfectly okay not to know. We are reasoning and science-minded people, and we are always exploring and questioning. Transhumanists know our species and our universe is in flux. In this way, our truths are always partially veiled—something that can change and something we might never catch up to completely. Transhumanists can get better at describing that truth—and closer at knowing it, too. But it will always be ahead of us, somewhere in the distance.

In the past, historical movements, like communism, environmentalism, or anarcho-primitivism have been steeped in cultures that were very single-minded. Their missions were to push their agendas forward while preserving their culture, values, and ideals. However, with a culture based on knowing it can never have a permanent culture, transhumanism is indeed in a strange conundrum.

Despite this, transhumanism aims to appeal to literally everyone. It aims to be a universal movement, regardless of the fact that many people have voiced issues against it. Luckily, many of those issues are fear or psychologically based, rather than true, rational objections. This is because transhumanism may be the least bigoted philosophy on Earth. Its central goal is to improve every person's life, and it makes no judgment on who those people are. Transhumanists are people of all races, abilities, religions, backgrounds, genders, employments, sexual orientations, educations, and wealth levels. It just, simply, wants to improve people's lives.

Unfortunately, we live in a world where "culture" has indoctrinated us to even go against our own health and well-being, or our children's health and well-being. Religious indoctrination is a key example—teaching us to handle rattlesnakes without fear of being bitten, forcing genital mutilation, and marrying young children off. But there are other forms of culture that also work against us with no religious bent—such as when people refuse vaccines, or when pregnant women wear high heels while gestating, or when people are trying to look hip riding motorcycles without helmets.

Our identity should not only be based on who we are, or where we come from, but also on where we're going and what we can become—especially in the 21st Century when science and technology is starting to change so many things about us. Transhumanists aim to make every person the very strongest, very best person they can be if they want to be. We aim to make it so that suffering, death, lack of ability, and lack of well-being never reach anyone ever again if they choose not to want it.

Transhumanists believe everyone has the possibility to grow greater than they are, and they encourage people to not remain in a culture of fear, deathism, irrationality, and even masochism that is ultimately self-defeating. Everyone on the planet can use improvement. Transhumanists are leading the charge to bring that improvement to the world and themselves.

****

CHAPTER II: EARLY WRITINGS

10) When Does Hindering Life Extension Science Become a Crime?

Every human being has both a minimum and a maximum amount of life hours left to live. If you add together the possible maximum life hours of every living person on the planet, you arrive at a special number: the optimum amount of time for our species to evolve, find happiness, and become the most that it can be. Many reasonable people feel we should attempt to achieve this maximum number of life hours for humankind. After all, very few people actually wish to prematurely die or wish for their fellow humans' premature deaths.

In a free and functioning democratic society, it's the duty of our leaders and government to implement laws and social strategies to maximize these life hours that we want to safeguard. Regardless of ideological, political, religious, or cultural beliefs, we expect our leaders and government to protect our lives and ensure the maximum length of our lifespans. Any other behavior cuts short the time human beings have left to live. Anything else becomes a crime of prematurely ending human lives. Anything else fits the common legal term we have for that type of reprehensible behavior: criminal manslaughter.

In 2001, former President George W. Bush restricted federal funding for stem cell research, one of the most promising fields of medicine in the 21st Century. Stem cells can be used to help fight disease and, therefore, can lengthen lives. Bush restricted the funding because his conservative religious beliefs—some stem cells came from aborted fetuses—conflicted with his fiduciary duty of helping millions of ailing, disease-stricken human beings. Much medical research in the United States relies heavily on government funding and the legal right to do the research. Ultimately, when a disapproving President limits public resources for a specific field of science, the research in that field slows down dramatically—even if that research would obviously lengthen and improve the lives of millions.

It's not just politicians that are prematurely ending our lives with what can be called "pro-death" policies and ideologies. In 2009, on a trip to Africa,

Pope Benedict XVI told journalists that the epidemic of AIDS would be worsened by encouraging people to use condoms. More than 25 million people have died from AIDS since the first cases began being reported in the news in the early 1980s. In numerous studies, condoms have been shown to help stop the spread of HIV, the virus that causes AIDS. This makes condoms one of the simplest and most affordable life extension tools on the planet. Unfathomably, the billion-person strong Catholic Church actively supports the idea that condom usage is sinful, despite the fact that such a malicious policy has helped sicken and kill a staggering amount of innocent people.

Hank Pellissier, a futurist and organizer of the conference Transhuman Visions, says, "The public majority disapproves of Christian Scientist and Jehovah's Witness parents who deny medicine to children afflicted with life-threatening illness. The public regards the anti-science attitudes of these faiths as unacceptable. Likewise, we should similarly disapprove of the withholding of any medicine or life extension practices that deter death for individuals, of any age."

Regrettably, in 2014, America continues to be permeated with an anti-life extension culture. Genetic engineering experiments in humans often have to pass numerous red-tape-laden government regulatory bodies in order to conduct any tests at all, especially at publically funded universities and research centers. Additionally, many states still ban human reproductive cloning, which could one day play a critical part in extending human life. The current US administration is also culpable. The White House is simply not doing enough to extend American lifespans. The US Government spends just 2% of the national budget on science and medical research, while their defense budget is over 20%, according to a 2011 US Office of Management Budget chart. Does President Obama not care about this fact, or is he unaware that not actively funding and supporting life extension research indeed shortens lives?

In my philosophical novel *The Transhumanist Wager*, there is a scene which takes place outside of a California courthouse where transhumanist activists are holding up a banner. The words inscribed on the banner sum up some eye-opening data:

By not actively funding life extension research, the amount of life hours the United States Government is stealing from its citizens is thousands of times more than all the American life hours lost in the Twin Towers tragedy, the AIDS epidemic, and the Vietnam War combined. Demand that your government federally fund transhuman research, nullify anti-science laws, and promote a life extension culture. The average human body can be made to live healthily and productively beyond age 150.

Some longevity experts think that with a small amount of funding—$50 billion dollars—targeted specifically towards life extension research and ending human mortality, average human lifespans could be increased by 25-50 years in about a decade's time. The world's net worth is over $200 trillion dollars, so the species can easily spare a fraction of its wealth to gain some of the most valuable commodities humans have: health and time.

Unfortunately, our species has already lost a massive amount of life hours; billions of lives have been unnecessarily cut short in the last 50 years because of widespread anti-science attitudes and policies. Even in the modern 21st Century, our evolutionary development continues to be significantly hampered by world leaders and governments who believe in non-empirical, faith-driven religious doctrines—most of which require the worship of deities whose teachings totally negate the need for radical life extension science. Virtually every major leader on the planet believes their "God" will give them an afterlife in a heavenly paradise, so living longer on planet Earth is just not that important.

Back in the real world, 150,000 people died yesterday. Another 150,000 will cease to exist today, and the same amount will disappear tomorrow. A good way to reverse this widespread deathist attitude should start with investigative government and non-government commissions examining whether public fiduciary duty requires acting in the best interest of people's health and longevity. Furthermore, investigative commissions should be set up to examine whether former and current top politicians and religious leaders are guilty of shortening people's lives for their own selfish beliefs and ideologies. Organizations and other global leaders that have done the same should be scrutinized and investigated too. And if fault or crimes against humanity are found, justice should be administered. After all, it's possible that the Catholic Church's stance on condoms will be responsible for more deaths in Africa than the Holocaust was responsible for in Europe. Over one million AIDS victims died in Africa last year alone. Catholicism is growing quickly in Africa, and there will soon be nearly 200 million Catholics on the continent.

As a civilization of advanced beings who desire to live longer, better, and more successfully, it is our responsibility to put government, religious institutions, big business, and other entities that endorse pro-death policies on notice. Society should stand ready to prosecute anyone that deliberately promotes agendas and actions that prematurely end people's useful lives. Stifling or hindering life extension science, education, and practices needs to be recognized as a legitimate crime.

****

11) Cryonics, Special Needs People, and the Coming Transhumanist Future

Recently, I was at Peet's Coffee writing an article on my laptop. A tired father walked into the shop with his adult son, a portly-looking 20-year-old weighing over 200 pounds. The son had Down syndrome, and his mental state was so confused that the father had to walk closely behind him, holding both of his shoulders to guide him. The son moaned as he walked, jerking forward in sharp, uncoordinated movements. Saliva bubbled out of his mouth.

I'm the parent of two young children (a 3-year-old and an 11-week-old infant), and my sympathy immediately went out to this father and his grueling burden in life. For many parents—especially a nonreligious one like myself—having an extreme special needs child is a daunting worry. Every five minutes in America, a child is born mentally retarded. That's over 100,000 kids a year. Approximately three percent of the American population has some form of severe cognitive dysfunction.

I watched the father place his order with the Peet's barista, receive his coffee, and lead his son to the condiment bar right next to me. The father released his son for a moment while he put creamer into his coffee. Within two seconds, the son arbitrarily lunged for my tea, spilling it all over my computer. He then proceeded to the next table and did the same with their drinks, yelling and grunting riotously.

Many people, including myself, jumped up and helped the father regain control of his son. It took only one look at the father's moist eyes to see how difficult this man's life was—filled with endless public apologies for his son's unpredictable behavior.

The question society must ask itself in the 21st Century—the age of transhumanist science and technology: genetic engineering, cyborgism, artificial intelligence, robotics, and radical life extension research—is what is the best way to handle such extreme special needs people? There are over seven million people in America with severe learning and cognitive disabilities—the most common are Down syndrome, Velocariofacial syndrome, and Fetal Alcohol syndrome—of which about 250,000 are institutionalized. Many of them are in poor health and will die prematurely. For both parents and society, the obligation is both massive and challenging. It costs many billions of dollars to keep extreme special needs people alive and not hurting themselves or others.

The field of cryonics—where human bodies are frozen using ultra-cold temperatures—has come a long way since the first person was preserved in 1967. Organizations like Alcor, Cryonics Institute, and Suspended Animation Inc. have together frozen ("suspended" in cryo-talk) a few hundred people. Eventually, science will figure out a way to bring them back to life—to revive them. Already, some people have survived death in freezing water for over an hour and have been brought back to life. Additionally, new techniques using a saline-cooling procedure can help restart the lives of people who have been recently declared clinically dead. Each year science advances, and the chances for reanimation of cryonic patients improve.

Given the feverish pace of scientific growth and innovation in the modern world, would it not be better to cryogenically freeze severely mentally retarded people with the hope of bringing them back to an age where science can cure them of their imperfection? More so, is it moral in the 21st Century to allow them to exist and die as they are, when likely in a matter of decades science will have what it needs to genetically alter them and make them cognitively normal? Don't we owe them the chance to be like us?

As a transhumanist philosopher, I advocate going further than just preserving special needs people after they die. I believe parents should have the legal right to painlessly put their extreme special needs children into a cryogenic state while they are biologically healthy and have years left on their lives (a process called cryothanasia). Some extreme special needs people are clearly unhappy, living in a nightmarish rollercoaster mental state—one that is also excruciatingly painful and crushing for their families. The all-important question to ask is: If it was you in their position—either as the parent or the special needs person—what would you want? The answers, at least for the nonreligious, are quite obvious.

So why then is this act illegal? Why is society afraid of evolving its perspective on this? Is it religion? Cultural stigma? Or are we simply lazy and prefer turning a blind eye to the controversial matter?

Many may say cryogenically preserving someone while they're still biologically healthy is murder (since it would technically involve stopping their life to successfully complete the process), but what do you call a person persistent on enslaving someone in decades-long confusion, insanity, and possible suffering when a more reasonable option exists? More importantly, like many other transhumanists and life extensionists, I no longer believe in death when it involves cryonics. Cryonics is more similar to sleeping or hibernation: a machine with the power button temporarily switched to "off." This is the 21st Century; dying is going the way of the dinosaurs. If you don't believe it, you're not reading scientific and medical journals. Additionally, you're not talking to trauma surgeons who can preserve technically deceased gun-shot victims for hours at a time before bringing them back to life.

Human civilization is at the cusp of achieving indefinite life extension for our species. Many leading bio-gerontologists say it's only a matter of decades before we can stop or reverse aging in people. Experiments are already succeeding with this in mice. Furthermore, hundreds of millions of dollars are being poured into genetic engineering research. Additionally, replacing old body parts with new artificial body parts will become commonplace in five to ten years. Perhaps most immediately promising, the use of stem cells to rehabilitate disease and malfunction in the brain is already being used with some success in research laboratories and hospitals. Clearly, if we can just get extreme special needs people to live long enough—or we can cryopreserve them if parents prefer—we will have a chance in the future to make them cognitively normal.

Currently, the situation today with extreme special needs people is anything but *normal*. While some are institutionalized and cared for by the state, many others are not. Some families make the choice to care for their special needs members. This is, of course, incredibly difficult to do and often leaves everyone miserable. Besides enormous time and financial loss, there are immeasurable emotional tolls. Marriages sometimes break up over attempting to provide the care. Healthy and intelligent siblings are regularly given the cold shoulder due to the constant demands of special needs siblings. Attempting even the most basic public outings with a special needs person (such as getting coffee at Peet's) can become a dangerous, complicated ordeal. The list of negative repercussions for anyone trying to provide care for a mentally retarded person goes on and on.

One of the main reasons I advocate cryonics as a possible consideration for severe special needs people—whether they're in the middle of their life or the end of it—is for the parent's sakes. Wouldn't parents rather live happy, productive, and liberated lives rather than spending their time changing diapers and spoon feeding an unruly adult with 20 more years to live? And by the way, that extra 20 years is actually going to be an extra 50 years in a decade's time given how fast life extension science is advancing.

Another point to consider are the financial aspects of cryonics for severe cases of special needs people. Cryogenically preserving someone costs approximately $150,000, and then approximately $1000 annually in maintenance and storage fees. Caring for a special needs person runs at least $1,000,000 over their lifetime according to a US Government CDC report in 2003, and that figure is likely much higher in severe cases, especially considering the increasingly length of lifespans due to modern medicine. (Of course, one also needs to add in 11 years of inflation from when that 2003 CDC report was published too.) In short, it's probably dozens of times cheaper to go the cryonics route.

To honor society's commitment to special needs people that are cryogenically preserved, we could put all the resources we were going to spend on their lifetime care into cryonic rejuvenation science and technology, as well as genetic engineering which in the future will likely be able to reverse mental retardation. That amount of redirected money would equal many billions of dollars. In return, such newly funded science would also help future-born people with special needs, as well as the human population as a whole. Crossover science would certainly occur and spur new technologies, medicines, jobs, and ideas.

The future is coming far more quickly than most people realize. Bionic arms now connect to human nervous systems. Computer chips are already being put in people's heads for medical reasons. Video games can be played using just brainwaves. Genetically engineering humans will become a reality in just a few years. Artificial general intelligence, the holy grail of technology that may solve many of humanity's problems, will arrive within a decade or two. With such incredible technology and science at our species' disposal, an entire new set of rights and wrongs, as well as moral ambiguities, will challenge us all.

As I mentioned before, the most pertinent question one can ask when facing such radical transhumanist technology is this: If it were me or if it was my child that could benefit from such advances and ideas, would I endorse it? In the case of the young man in Peet's afflicted with severe Down syndrome, who likely has decades left of his life, the controversial proposition of cryonic suspension should be considered. Anything else for him or the hundreds of thousands of others like him represents missed opportunity, possible injustice, and maybe even profound inhumanity. As a society living in the 21st Century, we can do better for the special needs people that deserve a chance at becoming normal.

(My view on parts of the content in this essay is changing some, as technology improves faster than anticipated. There may be opportunity now in as little as 10-25 years to improve the disabled mind via brainwave technology and other innovation. This means it might be more sensible for society to try to keep a disabled mind alive and operating as long as possible—even if only marginally or in severe pain—until a cure is found and applied.)

******

12) Despite Skepticism, Many People May Embrace Radical Transhumanist Technology in a Futurization of Values

As long as they're earthbound, most people shrug off the idea of being anything other than a biological human. Some people are even repulsed or angered by the concept of scientifically tampering with the human body and brain too much. However, the time is coming when radical technology will allow us to expand and significantly improve the abilities of our minds and the forms of our bodies. A transhumanist age is nearly upon the human race — an age where cyborgs, sentient robots, virtual lives based in computers and dramatically altered human beings may become commonplace.

Already, there are hundreds of universities, laboratories and companies around the world where transhumanist projects are underway. A transhumanist is a person who aims to move beyond the human being via science and technology. Some of the most eye-opening projects are military-oriented, such as the "Iron Man" armor suit being created for American soldiers. Trials runs of the suit are tentatively scheduled for this summer. Another well-known project is at Chalmers University of Technology in Sweden where scientists are connecting robotic limbs to the human nervous system of amputees, essentially creating cyborg-like people. The first arm surgeries are scheduled to occur in less than 12 months. Of course, private companies like Google are also very much involved in the broad field of transhumanism. They are spending many millions of dollars on creating artificial intelligence, which one day may have its own sentience and be thousands of times smarter than humans.

Even though some of these technologies seem frightening to the layperson, they should be here in a matter of years, not decades. One of the most exciting and controversial ideas of transhumanism is the complete integration of the human mind with a machine. Similar to the extraordinary technology featured in the movie The Matrix, humans may be able to download themselves into computers and live virtual existences.

Lately, I have been speaking more frequently about mind uploading in conferences, interviews and in casual conversations with friends. I often get asked in a highly dubious way: Could you really just let yourself disappear into a machine, Zoltan?

The study of how and why human beings and society accept technology and innovation is fascinating. Generally, people are wired to be wary and afraid of treading new paths and considering unknown ideas; we are engrained with a powerful "flight" mechanism, designed to preserve our safety and well-being. Yet, that has hardly stopped civilization from progress. The first time fire was seen by our homo erectus ancestors, it was likely treated as a great evil or a monster. Later, it became our species' foundation for warmth, disease-free food and light. The history of anesthesia is similar. At first, some considered it too unnatural before realizing how useful it was for successful surgery and medicine. Even the automobile was considered too loud and problematic when it first came out. Nonetheless, like all great technologies, society did embrace it, even if skeptically at first.

In time, many humans will also come to view mind uploading and virtual lives as just as important and real as biological human lives. Already on sites like Kickstarter, there are companies looking for funding that will create thought-capturing headsets and haptic feedback suits to bring us that much closer to complete virtual world immersion. Even virtual sex, considered bizarre by most, will likely come to be a popular way to enjoy intimacy with a partner. In an increasingly busy world where many travel for work and are away from loved ones for days at a time, such intimacy may be welcomed. Some may laugh at these concepts now, but the personal computer was laughed at by many too when it first came out.

A concept I've defined in my philosophical writings as the "futurization of values" promotes the idea that people should try to live according to where they believe they are going in life, and not only where they actually are. With science and technology advancing so rapidly, it would be valuable to begin examining the perspective from our projected future selves. In this way, we might not be so skeptical or afraid of new technology that might be beneficial to our species. Rather than mock and shrug off such advances that will soon be a part of our lives, we might consider instead what their value is and how they might improve our lives and those of our loved ones.

****

13) Origami Cranes: Who is Responsible for this Child's Death? (Introduction to the World's First Mainstream Media Column on Transhumanism: *Psychology Today's: The Transhumanist Philosopher*)

Colorful origami paper cranes appeared on a neighbor's front yard last week, as they often do on lawns across America when a child is dying from a brain tumor. The cranes are supposed to be a heartwarming symbol of eternity, life, and good luck, put up by family and friends to support that child.

Today that child died.

I live in a community of tree-lined streets in San Francisco where many kids go to college and pursue careers in technology, law, and medicine. It's a close-knit neighborhood with few issues. A sick child here usually gets the best healthcare possible. Two of the world's leading medical centers, UCSF and Stanford, are within close driving distance.

Unfortunately, when illness struck this 6-year-old child, the best medicine was not enough. Some people find it hard to believe in our modern world of smart phones and jet travel that we still can't stave off disease. Inevitably, everyone asks: Who or what is responsible for the death of this child? The answer is simple: *We* are all responsible.

Few people want to address the fact that science and medicine are lagging far behind where they could be if adequate resources were given to them. Even fewer people would agree that they are responsible for that fact. But make no mistake: We are all responsible. We are all responsible for the death of that child. We have not dedicated enough of our time, energy, and resources to the advancement of science and medicine. Furthermore, every time we give a dollar to a religious institution instead of to a scientific institution, every time we endorse a politician who cares more about lobbyists than our fumbling national education system, and every time we support our government's trillion dollar wars instead of a trillion dollar war on cancer, heart disease and diabetes, we support the premature death of innocent people.

I recently took a tour through the research center at SENS (Strategies for Engineered Negligible Senescence), one of the most prominent nonprofit scientific organizations attempting to stop aging and disease. Filled with white-gowned men and women bent over microscopes, SENS has many promising scientists in its labs. I asked Dr. Aubrey de Grey, Chief Science Officer of SENS, and one of the most visible anti-aging advocates in the world, what the SENS budget was for 2013.

"Less than five million dollars," he told me.

I gasped. That's tiny, I thought. Many nonprofit organizations like World Wildlife Fund, Feed the Children, and Red Cross have at least twenty-five times that amount for their annual budgets. And those organizations are not trying to stop the human race's most significant problem: dying.

There are many excellent research groups around the world trying to eliminate aging and disease, but all of them need more resources to speedily tackle the complexity of human longevity. Yet, nobody is giving enough money to these scientists because nobody cares enough. The US Government isn't doing much either to extend American lifespans. They spend just 2% of the national budget on science and medical research, while their defense budget is over 20%, according to a 2011 US Office of Management Budget chart.

What people don't realize is that with enough research money properly focused—$50 billion dollars, some experts say (the world's wealth is over $200 trillion)—human aging and the terror of disease can likely be halted. In the end of the day, controlling aging and disease are just more science puzzles waiting for the modern world to solve. We could help that process along if we changed the psychology of civilization's culture—a culture that largely believes human death is unstoppable and inevitable. Aging should be seen as a fixable problem, not as a destiny. The human race can overcome its biggest natural hurdle.

My new *Psychology Today* blog is titled *The Transhumanist Philosopher.* Every few weeks, I will be writing about how individuals and society are being transformed through rapidly advancing science and technology. I will be bringing you stories that dive into philosophical, sociological, and psychological perspectives of human enhancement, longevity issues, and transhumanism. I will be interviewing futurist and science leaders. I will be reviewing their books and projects. I will be exploring the philosophy of ending human aging and embracing indefinite lifespans.

One theme in my blog will always remain prominent. If, as a society, we choose to begin spending our energy and resources on health and longevity, then we will soon achieve the promise that life extension research, transhumanism, and human enhancement can bring us. We will soon become all that the human being is capable of becoming.

My goal of this blog is to quicken the coming of the day when there will be no more origami paper cranes on anyone's lawns ever again.

****

14) Transhumanists Frown on Talk of Genetic Engineering Moratorium

A wave of ethical discussions and admonitions recently appeared after Chinese scientists reported successfully editing an embryo's DNA. Many large media outlets interviewed scientists around the globe who chimed in on the ethical implications of the science, which ultimately could lead to designer babies. Some prominent scientists and bioethicists raised loud concerns about how genetic engineering might affect the future of humanity, and some of them went so far as to advocate for a moratorium on the research and technology.

Scientists and medical doctors at the US Transhumanist Party—a political organization I founded that prides itself on advancing science—condemned calls for any moratorium, saying such talk is anti-progress and anti-innovation. However, Transhumanist Party scientists did agree that an open and thorough discussion on the matter is warranted.

"We are at a time in human history when such radical genetic engineering technology may help eliminate disease and give us healthier children," said Dr. Lisa M. Memmel, a researcher, board-certified Obstetrician & Gynecologist, and Treasurer of the Transhumanist Party. "We should proceed with caution and an open discussion, but let's not derail the train before it has left the station."

Historically, scientific moratoriums have done little to stop progress, anyway. What they usually end up doing is giving other, less visible researchers and entrepreneurs the upper hand. When George W. Bush restricted federal funding for stem cell research in the United States—something he did primarily because of religious convictions—the research went overseas. Inevitably, other countries pursuing the science gained a notable advantage.

Jose Cordeiro, MBA, PhD, who is a faculty professor at Singularity University and the Technology Advisor of the US Transhumanist Party says, "Genetic manipulation of embryos should continue within ethical boundaries, and if the USA does not do it, other countries will do it."

Transhumanism, which literally means beyond human, is a growing international social movement that advocates for using science and technology to radically enhance the human being. Some of these technologies include robotic hearts, cranial implants, artificial limbs, exoskeleton suits, artificial intelligence, anti-aging research, and of course, the burgeoning field of genetic engineering.

Many major scientific breakthroughs and advances in the last few centuries have been challenged by luddites and anti-science naysayers, who usually warned that ethics were being negated or outright broken. But human ethics, at least when it concerns science, is bound to the idea of helping the human being to live better. Genetic engineering may eventually result in the elimination of heart disease, cancer, negative hereditary disorders, and even the ability to be affected by the flu. It also could make the next generation of human beings stronger and healthier, giving parents more choice in what they might want out of a child, including eye color, height, gender, athletic skill, and intelligence.

Critics—many of them fundamentally religious—worry that genetic engineering will create a race of nonhuman beings who resemble monsters. Their fears are overblown and tied more to Hollywood horror movies than actual science. The far greater likelihood is that genetic engineering will create a populace free of diseases and ailments that have plagued humanity for tens of thousands of years. In fact, genetic engineering could change the very nature of healthcare.

"It could revolutionize many if not all the ways human beings are cared for," says Dr. Joseph N. Carey, a plastic and reconstructive surgeon at the University of Southern California and the Medical Advisor to the US Transhumanist Party.

The one issue that many scientists and transhumanists do worry about with genetic engineering is whether only the rich will be able to afford such designer baby technology. Many transhumanists are humanitarians at heart, and they will surely advocate for sharing such important advances with all people, such that humanity as a whole may improve and become healthier than it's ever been.

****

15) A World Future Society Conference Speech: Everyone Faces a Transhumanist Wager

Recently, I had the honor to give a speech at the World Futurist Society's conference in Orlando, Florida. The World Futurist Society is the largest nonprofit organization of its kind with over 25,000 members in nearly 100 countries. Its yearly conference is a mecca for thousands of futurists looking to hear the latest forward-looking news and ideas. Hundreds of speeches, workshops, panels, meet-the-author sessions, poster presentations, and luncheons occurred.

My own speech at the conference was loosely based on an essay I recently wrote titled *Everyone Faces a Transhumanist Wager*. I wanted to share a condensed version of the talk because it presents a fundamental dilemma every human being on the planet must confront. Here's the shortened speech:

Ladies and gentlemen, we have a problem. Each one of us has a problem. In fact, no matter where you go on the planet, no matter who you find, every single person on Earth has this same dire problem.

That problem is our mortality. That problem is called death.

The reason it's a problem is because human beings love life. We all love the precious chance of existence. Even in one's darkest psychological despair, or one's most exhausting hardship, or one's most catastrophic tragedy, the thing we call life is still always miraculous. We cherish life and we don't want to lose it or have it end.

But end it will. No matter how much we wish otherwise. The stark truth is right before our eyes—that nothing in today's world can save us from death. The obviousness of this overwhelms us every time we see a loved one or a friend whose body is lifeless, never to reach out, touch, and communicate with us again. Death is final.

The great irony for our species is that we don't just have this one problem—but two problems. The second problem is nearly as vicious as the first. The second problem is the fact that most people around the world are just not worried about the first problem—they're not worried about dying. They're either religious and have the supposed afterlife all worked out, or they just don't care, or they just don't think conquering human death is possible. Whatever people's reasons, they just don't see the first problem as serious enough to warrant immediate concern—especially in a meaningful, tangible way that makes them not die. And by not recognizing death as a problem, many people have no reason to attempt to defeat it.

I have made it a mission in my life to make people aware of these two problems. It is why I wrote my philosophical novel *The Transhumanist Wager*. The concept of the Transhumanist Wager in the book is simple. It explains that in the 21st Century—the age of unprecedented technological innovation—it is a betrayal of ourselves (and the potential of our best selves) to not tackle and solve our two most pressing problems using modern science. More importantly, my book explains how we can solve these two problems.

But first, some of you are asking: What is a transhumanist? What does such a person want? What are the main goals? Some people around the world still don't know what transhumanism means. When explaining the term to people, I find it easiest to use the Latin translation. "Transhuman" literally means beyond human.

Transhumanist goals are broad and varied, but mostly they revolve around human beings using science and technology to radically improve and enhance themselves, their lives, and society. Transhumanists often concentrate on stopping or reversing aging—we are sometimes called life-extensionists or longevity advocates. Many transhumanists also focus on robotics, bionics, artificial intelligence, biohacking, and other similar fields of study. Transhumanists are often, but not always, nonreligious. They find meaning in their own lives and possibilities, without a divine creator. The philosophies of transhumanism make it possible that in the future—using extreme science and technology—one may become a so-called divine creator if they wanted. In almost all circumstances, transhumanists prefer reason over any other method of understanding to guide themselves in life.

Every transhumanist comes to their own realization of why they feel they are a transhumanist. Each path is unique, personal, and totally different than another. I want to tell you briefly about my path. I was first introduced to transhumanism as a philosophy student attending Columbia University in New York City. For a class assignment, I was told to read a magazine article on some of the recent breakthroughs in cryonics. The article described a small but passionate group of scientists who believed that science and technology would be able to bring frozen patients back to life in the future if they were preserved properly. The article also discussed the transhumanism movement, which it described as a community of reason-based futurists who wanted to use science and technology to improve their lives and live indefinitely. I was deeply intrigued. I finished that article and wanted to know more. I spent the next ten years reading everything I could find on future technologies, human enhancement, and transhumanism.

However, it wasn't until I was in the jungles of the demilitarized zone of Vietnam as a journalist for the National Geographic Channel that I came to dedicate my life to the field of transhumanism—that I came to the powerful conviction that human life should be preserved indefinitely. While in the jungle filming Vietnamese bomb diggers searching the ground for unexploded ordinances to recover and sell,

I almost stepped on a partially unburied landmine. My guide pushed me out of the way, and I fell to within a foot of the mine. Tens of thousands have died from landmines in the DMZ in the last forty years, and I was lucky I was not one of them.

For me, nothing was ever the same again after that moment. The landmine incident permanently stamped into my mind how fragile the human body was—how precious our minutes alive on this planet really are. Upon returning to the Unites States, I began writing *The Transhumanist Wager.* The reason I tell you my personal story about becoming a transhumanist is that every one of us has their own story. But the two main problems we each face: death, and general apathy of death—and the choice we must make regarding them: a Transhumanist Wager—that is not just for some people. It is for every reasonable person in the world.

Indeed, in the quickly advancing 21st Century, making a Transhumanist Wager approaches us now as an ultimatum—the most challenging one we may ever face. Luckily, given how fast modern science is growing and changing our lives, making the wager is also the only reasonable option. If you love life, you will dedicate yourself to finding a way to preserve that life.

Transhumanists do not want to preserve their life via heaven-promising religions, false hopes, an unconscious mystic super spirituality, or otherwise. There are only rational ways transhumanists will do it: through the tools they can create with their own hands; through the reason their brains can muster; and through the conviction their being prompts of them by not wanting to die and disappear. To do otherwise in today's world is to remain irrational and, as my novel discusses, to be masochistic and even borderline suicidal. In a world where we have the technology to travel to Mars, where we can video chat on our cell phones to someone 10,000 miles away, or we can triple the lifespan of mice with biotechnology, it's our evolutionary destiny to significantly extend our lives and to be transhuman.

Once you have identified the human race's two main problems, and you understand that you each face a Transhumanist Wager, the question is: what to do? How can you solve these problems and make the right choice in the wager.

It's quite simple, really. The journey of the transhumanist requires no ritual, no prayer, and no spiritual sacrifice or payment. It requires only your ability to reason. Ask yourself how you can best dedicate yourself to a specific cause of transhumanism and its various fields: aging research, cyborgology, stem cell science, suspended animation, singularitarianism, genetic engineering, machine intelligence, or the dozens of other areas. Then do it. For some, this may mean going into science or technology as a new career. For others it will mean volunteering in transhuman groups that need support. For some it will mean going into politics and pushing for more science-friendly laws. For others, it will mean donating resources to scientific centers and struggling innovators. For some, it will mean creating transhumanist art and using it a vehicle to push for a more scientific-minded society. For others it will mean just talking with friends and family about why you think science and technology are the best drivers of civilization.

Whatever it is that one can do, be transhumanist-minded. Be a people that belongs to a bright, rational scientific future, not one dogged by the old ways of archaic institutions, apathy, fear, or primitivism. Be transhuman, and let us all embrace our evolutionary destiny and the joys of perfect health and being that science can help us reach.

****

CHAPTER III: PERSONAL NOTES

16) Should I Have Had My Cat Cryonically Preserved?

I recently made the agonizing decision to euthanize my cat Ollie, who I adopted 13 years before from the streets.

Ollie had barely eaten or drank anything for five days and was dying from kidney failure. The veterinarian told me Ollie would probably be dead in 24 hours and suggested euthanizing him, so that his death wasn't caused by choking or something horrible like that when other organs failed. I reluctantly agreed.

Pet euthanasia generally includes a heavy morphine-based sedative that peacefully knocks the animal out, followed by a heart stopper-chemical injection. We euthanized Ollie on his favorite couch in my home. While the process seemed painless and quick, it was absolutely heartbreaking for my family and me.

Days after the death, a number of transhumanist friends consoled me and told me of their own dealings with pet deaths. Since I'm a life extension advocate, I'm well-versed in procedures for dealing with (and avoiding) human death. But I didn't really know much about pets.

It turns out many transhumanists have already thought of these and some have even undergone cryonic procedures with their animals—the process where they cryogenically freeze their pets in hopes to resurrect them in the future when the technology becomes available. The Michigan-based Cryonics Institute has 120 frozen pets. Costs to freeze a cat there are about $5700 dollars.

I thought deeply about doing this with Ollie, but decided against it for a few reasons—reasons I hope I won't later regret in my life as the world and technology rapidly advances.

To begin with, I was a little late in the process with Ollie. It was already 24 hours after he died that I began considering cryonics for him, and, like humans, the cryonics process works best if it's begun within hours of death—especially to preserve the brain and its memories. Also, $5,700 is quite a chunk of cash—plus there are yearly maintenance costs. Additionally, my kids are already yearning for another pet, and my parents have had seven different pets so far in their lives.

With this mind, I even considered the cheaper preservation methods, where a bucket filled with formaldehyde, glutereldahyde, or some other solution is used to preserve the pet. Then one can just keep the body in their garage. In this procedure, at least much of the tissue, bones, and organs might be able to be salvaged in the future when trying to reanimate the animal. Some people even stuff their pet or freeze-dry them to keep them in their house, looking as if they were almost alive.

In the end, I passed on all these options and opted for a normal burial of Ollie in my backyard, which my young daughters and wife attended.

The truth is I tend to believe I'll be merged with AI in about 30-40 years—and soon entering the Singularity afterward—so the idea of loving a cat indefinitely seemed less tangible.

I also wondered if in the future, we'd be able—and maybe even obligated—to make our pets hyper-intelligent via cranial implant technology and radical genetics. Then the animal, like an adult offspring, becomes intelligent enough to make its own decisions. What if Ollie didn't want to live? Or be so intelligent? Or even be my pet anymore? Such is the weird world of transhumanist thinking—and the future many of us will face in the coming decades.

Either way, Ollie's death started me down exploring the road of technology and science we are going to impose on the creatures we love. It turns out the pet industry is exploding with transhuman—or if you will, transanimal—themes. Most of these have nothing to do with death, but instead have to do with giving animals a better life so humans can enjoy them more.

For starters, an entire cottage industry on pet-tech wearables has emerged, with numerous start-ups already competing in the space. *Vice Motherboard* reported there will likely be exhibition space specifically dedicated at CES 2017 to pet tech. Currently, the leading wearables are Fitbit-like devices that help monitor dogs' whereabouts and health.

Of course, pets have long had RFID chip implants to help locate them, and their success has led the way of implants into humans—such as the one I now have in my hand. But the future of tech for pets is also developing too. There are devices like TailTalk and the KYON collar that can supposedly tell you about your animal's mood. Some companies have even launched projects to try to directly read the brain waves of pets, so one day you might be able to discuss Plato's Allegory of a Cave—or the adventures of Garfield.

As cool as some of the tech coming out for pets is, the world is headed for a massive transformation about how and what it wants in its future pets. CRISPR gene editing is already here, and the idea of creating a pet dinosaur is no longer a pipe dream. In fact, *MIT Technology Review* reports that Chinese scientists have already created "designer pets."

It's possible in just a few years time we will be creating new creatures that contain the very best elements pets have. Shed-less dogs. Uber-cuddly cats. Melodic singing songbirds. Why not combine them? Why not add some reptilian genes too, for excitement? In fact, why not just make a make a mini-Brontosaurus?

Of course, the other type of future pet will be created by secretive company Magic Leap, where sensors on your ceiling can put out a holographic pet image that you can interact with and order around. Why not have an eight-foot tall Tyrannosaurs Rex inside to scare off burglars when they break in? Or a 30-foot anaconda? Or a pack of wolves? Best yet, you can program the holographic wolves to take turns reading your toddlers The Three Little Pigs.

The future of transhuman pets, though, is not holographic or biological. It's robotic. The field of robotic dogs already available on the market is massive. There are a few dozen companies and types of robotic dogs out there. Some of these machines are designed to be legitimate guard dogs, and can offer real security via movement tracking mechanisms and security software. In the near future, some will offer Skype abilities, so you can see through cameras in their eyes what's happening in your house—like if your child is playing with the stove. Other robot dogs will have built in fire alarms that can register smoke in a child's room or spot a poisonous spider in the dark crawling on a bed crib.

In probably just five years, robot dogs will be so sophisticated they will walk our children to school, carry our groceries for us from the car, and probably even have built in drone capabilities to fly. We'll program them to catch rats but not fight with the neighbor's cat. They won't need to be fed, they'll know how to recharge themselves, and gone forever will be the days of shoveling dog poo. And of course, they'll easily beat us in chess.

Some new pet robots have fake fur too. In the future, we can expect robotic pets to have non-shedding, clean smelling fur that is dirt resistant and looks just like a real pet. And the pet's bodies will be soft and padded, with heat creating capabilities to keep you warm at night when it sleeps with you.

Like so many other things technology is changing for the human race, the central role pets play in our lives will also change. The domestication of animals has evolved for thousands of years, but the next 25 years may end pet relationships as we know them. While I'm still a little unsure whether I should've cryo-preserved my cat, I think Ollie would've found it strange to be brought back to a world with chess-playing robot dogs, holographic wolves in the living room, and mini-Tyrannosaurus Rexs cruising around the backyard.

****

17) The New American Dream: Let the Robots Take our Jobs

Many of us wake in the mornings to a dreaded alarm clock. After breakfast, we jump into our cars, battle traffic, and start a tiring 9 to 5 at work. Then we come home, turn on the tube, sip a beverage, and mostly veg. We do that all week long, waiting for the weekend when we might actually get time to travel somewhere, enjoy a hobby, or complete a fun project. Then we repeat, and it's only broken up by our measly two-week vacation. The American Dream is not so much a pilgrimage anymore, but a well-greased hamster wheel. We have been cajoled into an economic system that needs to infinitely grow in order to feed itself and feel satiated.

In the transhumanist age we are now entering, the same philosophy of keeping up with the Joneses is increasingly becoming a less viable economic policy. And the robots and software applications the Joneses are building to take our jobs are simply not something we can or should attempt to compete against. We won't win.

However, as human beings, we can evolve and be happier and more fulfilled than we've ever been before. The key is a shift in our thinking—and in the value we place in the kind of work we want to do and how we enjoy free time.

The near-term socioeconomic forecasts are pretty startling when you look closely at them. *The Washington Post* recently ran an essay pointing out that 3.5 million truck drivers are poised to lose their jobs when the trucking industry becomes driverless (which could start happening in about two years, I believe). The trucking industry is just one of hundreds of sectors that could become significantly automated within a decade.

For example, *Business Insider* recently shared footage of a robot doing the work of a food server in China—the restaurant owner bought the robot for a mere $13,000. Home improvement giant Lowes already has robots on some of its floors. Some hotel check-ins are now automated too, eliminating the need for receptionists. Few if any jobs, including those of journalists, lawyers, doctors, and politicians, are totally protected from automation anymore.

Like many other people, you're probably asking yourself: How can this possibly be a good thing? The simple answer is: It will be a good thing only if we make it one. To begin with, there's no point in pretending society can avoid a future Universal Basic Income (UBI)—one that meets basic living standards—of some sort in America and around the world if robots or software take most of the jobs. I prefer leasing out America's $200 trillion dollars worth of federal land and natural resources (called a Federal Land Dividend) to provide a basic income. But other ideas—largely progressive or liberal ones—include income redistribution via taxes, increased welfare, or a mass guaranteed basic income plan tech and robot companies exclusively pay. Either way, without some type of basic income in the future, there will be mass revolutions that could end in a dystopian civilization—leading essentially to what experts call a societal collapse.

Those are the basic options, since history and common sense tells us it's impossible to maintain a peaceful, free society with the rich growing wealthier due to cheaper labor and the rest of society growing poorer due to lack of income. Since I do believe that democracy generally works, I think people will eventually vote for the best interest of the majority—which will soon be comprised of the jobless on our current trajectory. I also tend to believe the rich of the world will want to see democracy thrive, something that has helped all classes and types of people prosper. The elite may not want to part with some of their money (I myself support many libertarian ideas) via wealth redistribution, but I think they probably want to avoid an ugly dystopian world even more—especially one where they would be despised rulers.

Of course, there are other, more radical alternatives, too. Susanne Tarkowski Tempelhof, Founder & CEO of Bitnation says, "To the contrary of what many people think, Basic Income doesn't have to be a socialist-leaning concept, it can be done on an entirely voluntary basis, through using the Bitcoin Blockchain technology, charging a fee on top of transactions, if people accept to pay that fee. Here at Bitnation we believe that as the world becomes more transhumanist, the nation state will loose relevance naturally—so avoiding national tax collection schemes in favor of voluntary contribution to online platforms is a more long term sustainable approach."

Whatever the future brings, it should also not be overlooked that the main goals when humanity creates technology is to gain freedom and prosperity. Broadly speaking, that is exactly what has occurred so far in history. Experts point out that technology is widely responsible for the positive progress the world has experienced since the Industrial Revolution. In the last few decades, that progress is even more pronounced. More people on planet Earth—regardless of wealth—are healthier, more educated, and living longer, according to a recent report from The World Bank. Just about every aspect of human experience has improved across the world as a result of technology. This will likely continue as further tech innovation occurs, even when—especially when—robots take our jobs.

As a political candidate, I am hoping to further this improved standard of living that is being experienced everywhere. I specifically advocate for some free education at all levels (including higher education) paid for by my Federal Land Dividend. In fact, I support increased education levels, too, including some forms of mandatory preschool and 4-year college for everyone.

After all, if people are living twice as long as before (most people born today will live at least 125 years, many experts say), education should also be lengthened. The thing with education is it gives people time to think about what they really want to do in life. And I don't mean how to better keep up with the Joneses and their robots—I mean discovering their own inner passion and skills. Maybe it's art. Maybe it's engineering. Maybe it's sports. Maybe it's science. Whatever it is, a longer, more in depth education gives a person's spirit and mind the proper environment to decide what its heart is all about.

In the future—with less work and responsibility due to robots taking our jobs and leaving us only to collect our UBI—we might find there is a lot more to life than buying the latest trinkets from Walmart, or zoning out late at night in front of a television, or worrying about how poorly our bosses treat us at work. I say let the robots come. They may take our jobs, but they bring us freedom as well. With that freedom, we can become the best human beings we are capable of—a people full of passion, education, and a newly discovered drive of what it means to be alive. Perhaps it's time to reimagine the American Dream.

****

18) Baggage Culture and Why Embracing Transhumanism Doesn't Come Easy

Twenty years ago, while in college and wondering why everyone else in the world wasn't hell-bent on trying to live indefinitely via the promising fields of transhumanist science, I began working on the idea of what mass culture is and if it was holding back people from wanting to maximize their lifespans and human potential. I came up with the concept baggage culture, which is explored in detail in my novel *The Transhumanist Wager* and its philosophy Teleological Egocentric Functionalism (TEF).

Upon the request of my friends at Movement for Indefinite Life Extension (MILE), I recently condensed my thoughts on baggage culture in my speech at the Brighter Brains Future of Emotional Health and Intelligence Conference at University of California, Berkeley. Here's a summary of that recent talk:

For many thousands of years now, the human race has been indoctrinated to submit to orthodoxy and to cower before authority, and to swallow endless nonsense from both. We have been brainwashed to sacrifice our innermost desires, our most obvious needs, our most natural outlook on reality, just to live as a hostage in a cage of carefully regulated and fabricated cognitive existence. Virtually everyone and everything—our countries, customs, faiths, leaders, relatives, friends, lifestyles, even our own memories—have been manipulating and pressuring us to shun fresh, unconventional thoughts. Especially transhuman-oriented thoughts. There has been a pervasive worldwide moratorium on thinking about what the human being is capable of and its possible evolutionary advancement in terms that make a substantial difference in reality.

Why has this happened? To transhumanists, the reason is obvious: We—the people of the world—have allowed it to happen. Each of us is guilty for not heeding a higher calling: a more logical, more ambitious, more sublime direction for our life, and a journey to our best self. Our great flaw is the mistaken way in which we choose to interpret existence; our subscription and obedience to the cultural constructs that government, organized religion, ethnic heritage, mega-corporations, and mass media have built around, and within, nearly every thought and action we make. Their web of indoctrination has wholly swamped our lives. Sadly, most of us don't even know this has happened. Most of us are living on this planet in utter delusion, conforming to a largely manufactured and forced reality.

Throughout our lives and modern history, civilization has erroneously subscribed to the vision that the human being is a marvelous, ingeniously assembled specimen of life: a work of divine creation and sweeping beauty, whose culture and intellect is profound like the cosmos itself. What a joke. The cruel truth is we are a frail, hacked-together organism living within a global culture of irrationality, pettiness, and deception. The specific reason our existing human culture is so malformed is that, throughout history, past cultural constructs of more primitive societies were not discarded as they became irrelevant or outdated. To survive, it was not evolutionarily required to rid ourselves of unnecessary idiosyncrasies and practiced customs—such as nonsensical superstitions, masochistic religiosity, and shackling morality—even though they were foolish to uphold. As a result, damaging, wasteful, and useless behavioral patterns were passed on both socially and individually from generation to generation.

So now, modern humans are a weighed-down species, burdened by cumbersome past rubbish that's mostly crudely stacked, obsolete cultural constructs through which our minds perceive reality. I call this baggage culture. And it's caused nearly all human life to be degenerate and apathetic compared to what it could be. Our species' mindset and powers of perception are currently too lumbering and unfit for what a sophisticated, nimble entity really needs of itself. Our lives are cursed because of the polluted cultural prism our thoughts must exist within and communicate through. In Sisyphean tragedy, we are doomed to grovel, to falter, to repeat our same pathetic mistakes, day after day, year after year, century after century. We need to transition from our defective culture into a new one that directly confronts these issues and sets our minds and transhuman possibilities free.

The twisted history of our baggage culture extends back many millennia. It started long ago with the inception of civilization, when charismatic leaders and ruling clans began forming permanent communities. Over time, these rulers learned they could preserve their platforms of power by controlling their communities' thinking and behavioral patterns. Their agendas were simple: dominate with fear through violence; stifle revolutionary and freethinking ambitions; teach adherence to leadership and community before self; implement forms of thought and behavioral control that encourage social cooperation and production, such as communal customs, prayers, taboos, and rites. Variations abounded, but these were the early convoluted versions of human culture and its main intent: to control. Henceforth, culture's core function became a means of forcing conformity, to transform the individual into a tool of submission and production for the ruling elite.

As generations passed, these rulers and their predecessors continually revised and enlarged their constructs of culture, force-feeding the functional and nonfunctional—rational and irrational—parts to our forbears. Naturally, it didn't take long in evolutionary terms before people everywhere existed within a universal baggage culture, full of compounded dysfunction. Of course, in modern times, control of human culture has changed hands from the ruling elite to whole governments, religious institutions, multi-continent ethnic groups, and most recently, to mega-corporations and mass media. As the complexities and population of the world ballooned, baggage culture continued to prove versatile and useful to whatever cause it engaged. Nations governed through it. Religions preached through it. Ethnic groups taught their heritages through it. Big business sold through it. And the media communicated through it.

To cement their authoritarian agendas, these supersized institutions' advancing baggage culture implemented ever more effective methods of control over society. Chief and most potent amongst them was the inversion of reason, where cultural forces obliged us to rationally accept the irrational. By corrupting the rational way we thought and interpreted life, they simultaneously corrupted the necessity and power of reason altogether. In that devious way, mysticism, ancestral divinity, the supernatural, religion, and even the institutions' all-important puffed-up selves were seen as valid outcomes of a supposedly sensible, straightforward, and successful society.

Among many others, altruism, filial piety, and consumer addiction to unnecessary materialism were other methods of control. However, to transhumanists, the most grotesque of all the methods was the perpetuation of fear in our lives; not by the threat of violence, but by implicit guilt. This powerful psychological addiction of worrying about what others think of us, and about what is socially acceptable to others, has been systematically instilled in humans for thousands of years, perpetrated by every world religion, ethnicity, and government. Its aim is to weaken people's wills and to silence their most precious independent tool: the ability to freely, guiltlessly, and publicly judge and criticize the world around them. In that way, people became afraid to pick apart others and their behaviors; afraid to deride society and its routines; afraid to upend their own world and circumstances; and, ultimately, afraid to differentiate between good and evil, utility and irrationality, strength and weakness, equal and non-equal—essentially all value itself. Such pervasive social control through the fear of others' opinions has left us meek, ashamed, and largely unwilling to openly question or challenge a thing like the omnipresent state. Or our sacred heritages. Or the rife sense of needing to be wealthier than our neighbors. Or our supposedly sinless and perfect gods. The spicy, troublesome, confrontational bigot in us is often our best and most useful part, and they have strangled it out of most of us in the guise of what they call "open-mindedness" or "politically correct social behavior."

Ultimately, implicit guilt and culture's many other devices of submission are designed to make us totally subscribe to one single concept: we should be afraid to rise to being as powerful an entity as we can; we should be afraid to try to become an omnipotent God. That is the essence and outcome of our baggage culture.

The truth is so simple to see once we understand it: Religion, ethnic heritage, state power, material addiction, and media entrapment are nothing more than pieces of an intangible psychological construct designed to keep us thinking and living a certain way. It's designed to keep us in fear of becoming as powerful as we can be; to keep us producing for others and contributing to their overall gain, and not our own.

Today, our species' baggage culture is a gargantuan mindless monster, consuming and dominating everything it can. Even its main pushers—the overarching institutions—can't control it anymore; instead, they always find it controlling and devouring them. There's no escape from the confusion and redundancy anymore, from the vestigial aspects of stacking useless cultural constructs upon each other. If you think one tailbone in the human body is pointless, imagine a hundred of them weighing you down. Figuratively, that's what baggage culture looks like. Many of our thoughts are piles of ignorance and erroneous ideas stacked upon piles of ignorance and erroneous ideas. We are unable to think freely and escape our slovenly, derelict pasts.

This, sadly, is baggage culture. And it's the primary reason we don't demand more of our lives and of our possible transhumanist future.

****

19) Should Surfing Be Allowed During the Pandemic?

Many surfers like me believe that surfing is more than just a sport; we consider it a way of life. Being in the ocean and riding waves can be ecstatic and spiritual.

But because of the coronavirus pandemic, many beaches and surf spots around the world are off limits. Citations have been issued for riding waves. On Instagram, you can watch a stand-up paddleboarder surfing Malibu alone for a blessed time before Los Angeles County Fire Lifeguards chase him down in a boat and arrest him.

I learned to surf in California when I was 10 years old in the early 1980s, and I still ride waves as often as possible. I usually go to a popular reef break called the Patch, which is a 30-minute drive from my home here in the Bay Area. The Patch is a softly peeling right-hander set among picturesque redwood-forested mountains, located just off the main beach in a reclusive hippie town called Bolinas. It's a place where day-tripping San Franciscans and locals hang out on the sand, smoke pot and build fires on the beach while their dogs frolic at the water's edge.

Once the coronavirus came, so did a Bolinas lockdown. Concerned citizens and government officials put up signs along the road into town, reading "Surfers Stay Home, Save Lives" and "Beaches

Closed Due to Covid-19." Some residents even stood along the road and yelled at cars with surfboards on their roofs to turn around.

I don't think many surfers obeyed. I didn't. I just couldn't see how walking out of my house, getting into my car, parking near the beach, and paddling into waves could be dangerous for anyone. Even on the beach — which hasn't been crowded since the pandemic hit — most people were wearing masks and practicing social distancing. In the water, we were always considerably more than six feet apart from one another.

The question in Bolinas — to surf or not to surf — is not just a local one. Many of the best waves in California, such as Steamer Lane in Santa Cruz, are walking distance from densely populated downtowns, complicating quarantine directives. This is also true of many surf spots worldwide.

Exercise is important, especially during stressful times. But many surfers, including myself, also use surfing as a form of healing and therapy. I spent a lot of time tearfully in the water after my father died three years ago. Like thousands of others, I consider surfing an essential activity.

Our governor, Gavin Newsom, and other state authorities don't seem to care much about surfing as a way of life. Up and down the coast, they have threatened surfers with $1,000 fines and closed off beaches, including one of the most famous surf spots, Trestles, in Orange County.

Similar crackdowns have occurred at other renowned spots around the world, including Mundaka in Spain, Surfers Paradise in Australia and Jeffreys Bay in South Africa.

A few days ago, a county sheriff's officer stood outside his vehicle in the parking area of the beach in Bolinas, waving off visitors and telling surfers to go home. Like many other surfers, I avoided him by parking on a side street. I suited up and after making sure he was looking the other way, sprinted to the water. I caught my first wave of the day a minute later.

I understand that quarantine rules must apply to everyone or the plan to flatten the curve doesn't work. But I doubt that surfing alone jeopardizes the health of society in any statistically meaningful way, especially because all the surfers I've seen have been careful to practice social distancing in and out of the water. The physical, mental and spiritual benefits to surfing outweigh the tiny chance a surfer might become infected or infect someone else.

I prefer how Hawaii, surfing's home, has handled the situation. While relaxing on the beach is forbidden, swimmers and surfers can go in the water so long as they stay six feet from one another. That policy seems fair and sensible.

A handful of closed beaches throughout the United States are now being reopened, with social distancing rules in place. But many remain off limits and are likely to stay that way at least through May, if not longer. I hope the authorities in those states and localities will consider allowing responsible surfers in the water.

****

20) Trojan Horse: Why I'm Running for President as a Republican

The speed of change we've seen in the last few decades is nothing compared to what's on the horizon with the emergence of radical new technology and science. Already, California has major companies working on brain implants that connect our thoughts in real time to the internet. Artificial Intelligence has become one of the most important military issues of our time. And designer babies born through genetic editing are already alive.

These innovations are part of the burgeoning field of transhumanism, the movement to enhance human bodies and lives with transformative technologies. I have dedicated my life to promoting transhumanism in a rational, objective, and fiscally conservative way. But transhumanism is under siege. Socialists, authoritarians, and the activist far left want it for their own. If allowed, they'll weaponize, stifle, and propagandize transhumanism until nobody wants this future because it's downright dystopian. Getting to a bright future of transhumanism requires capitalistic entrepreneurialism and a hands-off approach from the government.

For the sake of humankind's future — as well as a more nuanced meaning of American greatness — conservatives must steer the direction of transhumanism. Whoever owns transhumanism — whether you're a Republican or a Democrat, whether you're conservative or liberal, whether you're a democracy or a totalitarian regime — will own the future. Vladimir Putin has said as much.

There's not a moment to lose in this race. Silicon Valley — ground zero of transhumanism — is heading down a road of early socialism after formerly being quasi-libertarian. And authoritarian China is already leading American efficiency and innovation in many ways. Adding to all this is the left's continuous cries for more and more regulation. Only a free, deregulated America can catch up lost ground to China and Russia — and ensure that America retains its leadership as the world moves forward into a brave new future.

That new future could be what lasting American greatness is really all about. Greatness for America is still attainable, but it will require a massive leap forward, both culturally, intellectually, and scientifically to achieve it. It can do this by leading the world forward in using geo-engineering to control the Earth's atmosphere to mitigate climate change. It can create a universal basic income — without raising taxes — as it accepts that the great majority of human jobs will be lost to automation, robots and AI with 20 years' time. It can even use genetic editing to end disease and aging in its citizenry. This is what modern greatness might look like to a transhumanist.
Instead, our leaders are seeking to make America great by disrespecting democratic institutions, jarring longstanding international allies, and acting childishly on social media. America cannot be made great by pretending to bring back old jobs when China is on track to become the world's leading economy in just a few years' time via the creation of 65 million new jobs in the last five years alone.

Notwithstanding Trump's personal foibles, I was cautiously excited with the 2016 change of leadership at the top of America. I hoped the system would be jolted in a positive way. I even applied for a job in the current administration. But as time went on, the corruption of the White House, its political posturing, and the gridlock of Washington seem the same as ever. The national debt has hit a new record. Inequality is growing. Smoking a joint can still land you in jail. A new border wall, an economy surviving on irrationally free government money, or a questionable trade war are not recipes for greatness. Neither is bogging down the nation with impeachment proceedings.

Under the current administration, America's technological edge is slowly being handed on a platter to socialists and authoritarian powers. Even more important than Silicon Valley's dangerous liberalism is that China is increasingly outdoing us in science and tech — the stuff that historically has bettered everyday American lives more than anything else. The writing is on the wall. While the White House offers faux populism, a phony culture war, and regressive religious morality in the face of innovation, secular China is rushing forward to claim its mantle as the inevitable de facto world leader. Designer babies, artificial intelligence, and revolutionary green innovation — China arguably has a leading grip on all these markets.

I am unable to stand by and watch America fall short of its epic potential. It's for this reason that I am running for the US presidency — and running directly as a Republican against President Trump. I want to help ensure that America doesn't become second-best because of old guard politics, vacuous virtue signaling, and a lack of original thinking. I also want to open the Republican Party to 21st century ideas and social mores, which they need to accept to remain competitive against the Democrats and growing socialist base in America.

Raised in a Republican household by immigrant parents who believed in fiscal responsibility, family values, and hard work, I am coming back to the party of my father after being essentially an independent for a long time. While my father died a few years ago, my mom's 53-year marriage to him was a testament to loyalty and love, and the kind of values the Grand Old Party once possessed but seems recently to have eschewed. While I was the presidential nominee of a politically neutral, science-oriented party (Transhumanist Party), and was also an endorsed libertarian candidate in the 2018 California gubernatorial race, I have come to see the Republican Party as the best chance in 2020 to grab hold of enduring American values and achieve the aims the United States sorely needs.

It won't be easy for me. I represent a younger and more evolved type of Republican, forged more in the form of our revolutionary founders than the eroded Trumpian GOP. I'm fiscally conservative yet socially liberal. I'm religiously agnostic but am firm on the separation of church and state. I insist on shrinking government but realize its essential necessity for the greater welfare of our country.

At age 46, I've become a passionate advocate for sensible application of radical science and technology, consulting for governments and the US military. My public career was launched in 2013 when I published my novel *The Transhumanist Wager*, which is often compared to Ayn Rand's *Atlas Shrugged*. Rand's Objectivism was an important influence in my life, but it was my nearly five years as a journalist for National Geographic that defined my passion for politics.

As a journalist, I've seen horrors while on assignment in the 100+ countries I've visited — everything from genocide to the complete destruction of forests to gut-wrenching human rights abuses. I've covered many of those stories, from street children taking drugs to music helping overcome racism to fighter pilots losing their jobs to drones.

While my 20s were spent seeing and reporting on the world, my 30's were spent building various businesses; the most successful of which was in real estate development. People sometimes ask if I've worked with my hands. You betcha. I have personally built or renovated many homes in my life, using every power tool imaginable, doing every type of job. My hard work paid off; selling my business portfolio enabled me to stop formally working and dedicate myself full-time to writing and politics.

What most people know me for is helping to spearhead the transhumanist movement, which is surging into a bonafide mainstream phenomenon. Transhumanism may seem esoteric, but make no mistake, it is at the core of the world's most valuable companies like Apple, Google, and Microsoft. We are using more and more radical tech in our lives, and someday soon, a lot of that tech will be making its way into our bodies. I already have a tiny chip implant in my hand that I can text with, start a car, and open doors. The coming tech we'll see in the next five years will be more radical than what we've experienced in the last twenty. The new tech coming out of California's Bay Area, where I live with my physician wife and kids, will ultimately replace jobs — and maybe even biological human beings. A core part of my presidential campaign is not creating jobs but positioning America for the eventuality that there will be very few jobs left because robots will take nearly all of them.

It's for this reason that I support a universal basic income. My version doesn't raise taxes at all. Similar to Alaska, it utilizes and monetizes the 800 million federal acres of unused land America possesses. It's called a Federal Land Dividend and is more than enough to give every American $1,000 a month.

I also support spending less money on military conflicts, and more money on fighting the wars that really matter to Americans, like the war on cancer, heart disease, and Alzheimer's. Finally, a campaign agenda of mine is to legalize drugs, and spend the saved drug war money on rehabilitation and education. Let's get nonviolent criminals out of prison and back to work. And while we're at it, let's turn those now empty prisons into free public colleges.

The fact I'm pro-choice, pro-immigration, secular, and even support LGBTQ rights doesn't sit well with many Republicans, especially the older ones. They see me as a potential Trojan Horse. But there are plenty of younger Republicans like me that are fiscally conservative but also embrace a wide array of social freedoms. We aim to usher in humanity's radical new future in fiscally responsible way.

With authoritarian China looming evermore powerful on the horizon, Democrats increasingly embracing socialism, and robots poised to forever change our economies, a new vision of conservativism is needed. I aim to bring that into America's political sphere so that a new guard of Republicans might emerge and hold its ground against anti-American forces. I look forward to telling you more about it on the campaign trail, and how America might not just become great, but also the most innovative nation in the world again.

****

21) A Letter About Coronavirus, the Longevity Movement, & Why Quarantining is Killing Us

Dear Fellow Humans,

If you believe in the life extension movement of trying to live indefinitely through science and technology, then you likely should not support the worldwide quarantine (at least don't support it over 14 days in the West where we don't have the ability to do it as efficiently as Asia). It's horrible that so many lives will be lost by COVID-19, but in a "worse-case scenario" it's likely 100 million people (a more likely case is about a million people despite what sensational media tells you) will die globally (mostly older people who have only a few years left to live due to their underlying medical conditions of aging — and who have likely been kept alive due to science and 21st Century medicine anyway). But the damage we could cause (and almost certainly are causing) with the quarantine and shut down to the US and global economy may cost the life extension movement and its scientific research possibly three to five years of progress — because the funding, projects, and jobs around the anti-aging industry will disappear for a notable time. The math shows that if we achieve indefinite lifespans for the human race by the year 2035 vs 2040, approximately 250 million lives will be spared and could then go on indefinitely. The aging math (or life hours) for any transhumanist shows that if we care about human life and longevity — about how long people alive today live — then we should not quarantine the world right now, but get the economy going again as a first priority so that we may fund the future of anti-aging science for the species. Some of us call this reasoning the Transhumanist Wager. For the sake of everyone alive today, it must be acknowledged that there is a dramatically larger percent gain (many thousands of percent) of overall life years for our species by not quarantining and shutting down the world. This is all a horrible scenario, and one I am terribly sad to share with you, but that doesn't mean we should cower from facts. We owe our species the most courageous decision for its long-term longevity of all its living citizens.

Live long and prosper, Zoltan Istvan

****

22) Death Threats, Freedom, Transhumanism, and the Future

Last week, I published a guest post at *Wired UK* called *It's Time to Consider Restricting Human Breeding.* It was an opinion article that generated many commentary stories, over a thousand comments across the web, and even a few death threats for me.

Frankly, the article itself is quite gentle and exploratory. The intro reads: Given the number of children that starve each day, dwindling planetary resources, and the coming transhumanist era, it might be time to consider restricting human breeding, argues futurist Zoltan Istvan in this guest post.

The way the conservative press assaulted and twisted the article out-of-shape—everyone from *National Review* to *Life Site News* to *PJ Media* to *The American Catholic* to the Acton Institute—you'd think I had suggested we kill children, not save them from starvation and give them a better future, as the article seeks to do.

What disturbed me the most about the bitter commentaries was the misinterpreted idea of freedom by so many people, most who I'm assuming have not visited authoritarian regimes or reported on war zones like I have. Here in America, we live in a society that is filled with cherished liberties. We can visit most any beach and dig our toes in the sand whenever we want. We can go to the movies and be entertained. We can shop at Walmart and acquire the trinkets offered.

However, beyond the surface, those liberties are always limited. We can visit the beach, but swimming in the ocean may or may not be allowed depending on whether lifeguards have blackballed the water. At the movie theaters, we can only see what films are offered, and we can't yell "fire" while watching any of them. At Walmart, we'll need money to buy things or we'll be arrested if caught stealing.

No one has perfect freedom, no matter how much we'd like to pretend we do or how sacred we hold the concept. Freedom is always a question of how many liberties we possess and can actually exercise. Moreover, in a tediously litigious society such as America, losing or gaining any major freedoms is increasingly rare. Significant reforms of freedom in the 21st Century (and the very end of the 20th Century) have rarely occurred outside of technological necessity. This is because if you move noticeably outside the social and legally accepted norm, you're likely to get both sued and paraded around the world in social media (and possibly mass media too). Such is living in a democratic nation with over a million lawyers, 300 million cell phones, and a half billion social media accounts, like Facebook or Google+. The guarantees are no longer just death and taxes, but litigation and internet infamy. The world is increasingly being decided on twitter, and later in court if you're still not happy and have the time and money to pursue it.

Whether that's good news or not is up for debate, but beyond question is how difficult it has become to lose our democratic freedoms in the 21st Century, which continue to grow globally every year according to UN reports.

Adding to this inertia on the evolution of freedom is the ever-expanding Wall Street and the tight leash the Federal Reserve has it on. A significant portion of the country's net worth and success revolve around tradable security markets—and they do not withstand chaos or uncertainty well. Government, conglomerates, and those in power are always on careful watch to not see a swan dive of the market, as we did in 2009, where entire fortunes were lost, major companies disappeared, hundreds of thousands of families lost homes, and Occupy movements threatened civil stability. Neither the government nor its people can afford to gain too much freedom or power—or to give it away. Society and its modern civilization is a delicate balance of moving forward wearing the tight yoke of progress. Achieving prosperity involves sacrifice. That sacrifice is taken from your sense of freedom, like it or not.

Change is coming, though. In the next 25 years—the coming transhumanist era— technology will allow us to do things we thought we could only imagine before. The near future will be wrought with unique enterprises: downloading our minds into computers, using artificial wombs (ectogenesis) to give birth, and utilizing life extension medicine which will allow us to live past 150. Each one of those advances will significantly challenge the major social institutions—marriage, child rearing, and religion, to name just a few—that currently influence our lives.

To my conservative critics and death threat-toting commenters, that was also the point of my *Wired UK* article—that coming technology will chip away at institutions many people erroneously believe are forever. They are not forever, though. In fact, the earth is already shifting below our feet. With almost biblical fervor, science and technology are allowing quadriplegic people to walk. It's making the deaf hear. It's giving amputee soldiers limbs again. But that's just the beginning. In 10 years time, those bionic limbs will be better than any Olympic athlete weight lifter has. In 20 years time, half of your neighbors will likely have one—because they will be far better than organic limbs.

But advancement isn't always benign. Technology is also allowing people to have eight babies at once, such as the Octomom. In fact, eight babies at once is nothing. In the next ten years, someone might have 20 babies at once if they want, given how many eggs the ovaries can produce. Obviously, as technology changes, so should laws. And naturally, so should our concept of freedom.

To all those out there who subscribe to pursuing liberty, as I do, please take a look around and ask yourself if you really understand the nature of it. You'll likely find as I have, that freedom is partially a misnomer, largely constructed on what we make of it and the limits of our own vision and capacity. In the 21st Century, prepare to have concepts of freedom, privacy, identity, sexual orientation, ethnic heritage, political orientation, and spiritual perspectives dramatically challenged and forever changed.

Soon, we'll look in the sky and drones will drop off our pizzas. Implanted chips will monitor our children's whereabouts and their health. And satellite cameras will know our every move, every second of the day—and all of it will be saved in a data base for easy examination. Freedom. Not so much anymore. Privacy. Basically soon to be gone.

In the future, transhumanists, Artificial Intelligence machines, cyborg beings, hive mind entities, and digital avatars of ourselves may not even value freedom much. They may value cooperation and pursuit of advancement more, and the power and well-being that comes with that.

To the luddites and conservatives who are skeptical and afraid, I can only say: Try to get used to it. It's our future. It's evolution. A better version of freedom rests in embracing technology and science, and using it to advance the individual and the species' evolution. Freedom is no longer some high school kid being inspired by *Atlas Shrugged*. Freedom changes with time and evolution of technology. Don't hold it hostage to 20th Century philosophers, historical events, and ideals.

So the next time you read a controversial opinion article, instead of talking about "lynching" the author or what "gauge" shotgun you're going to use when you shoot him, maybe present a better idea to solve the problems he's trying to address. After all, my article was about this: Over 50 million kids starved to death on planet Earth in the last 30 years. And 15% of kids go hungry in America, the supposed wealthiest country in the world. Society can do better than that. Much better. Technology, an upgrade of our morality, and a more open-minded approach can help.

****

CHAPTER IV: CULTURAL PHILOSOPHY

23) How Brain Implants (and Other Technology) Could Make the Death Penalty Obsolete

The death penalty is one of America's most contentious issues. Critics complain that capital punishment is inhumane, pointing out how some executions have failed to quickly kill criminals (and instead tortured them). Supporters of the death penalty fire back saying capital punishment deters violent crime in society and serves justice to wronged victims. Complicating the matter is that political, ethnic, and religious lines don't easily distinguish death penalty advocates from its critics. In fact, only 31 states even allow capital punishment, so America is largely divided on the issue.

Regardless of the debate—which shows no signs of easing as we head into elections—I think technology will change the entire conversation in the next 10 to 20 years, rendering many of the most potent issues obsolete.

For example, it's likely we will have cranial implants in two decades time that will be able to send signals to our brains that manipulate our behaviors. Those implants will be able to control out-of-control tempers and violent actions—and maybe even unsavory thoughts. This type of tech raises the obvious question: Instead of killing someone who has committed a terrible crime, should we instead alter their brain and the way it functions to make them a better person?

Recently, the commercially available Thync device made headlines for being able to alter our moods. Additionally, nearly a half million people already have implants in their heads, most to overcome deafness, but some to help with Alzheimer's or epilepsy. So the technology to change behavior and alter the brain isn't science fiction. The science, in some ways, is already here—and certainly poised to grow, especially with Obama's $3 billion dollar BRAIN initiative, of which $70 million went to DARPA, partially for cranial implant research.

Some people may complain that implants are too invasive and extreme. But similar outcomes—especially in altering criminal's minds to better fit society's goals—may be accomplished by genetic engineering, nanotechnology, or even super drugs. In fact, many criminals are already given powerful drugs, which make them quite different that they might be without them. After all, some people—including myself—believe much violent crime is a version of mental disease.

With so much scientific possibility on the near-term horizon of changing someone's criminal behavior and attitudes, the real debate society may end up having soon is not whether to execute people, but whether society should advocate for cerebral reconditioning of criminals—in other words, a lobotomy.

Because I want to believe in the good of human beings, and I also think all human existence has some value, I'm on the lookout for ways to preserve life and maximize its usefulness in society.

One other method that could be considered for death row criminals is cryonics. The movie *Minority Report*, which features precogs who can see crime activity in the future, show other ways violent criminals are dealt with: namely a form of suspended animation where criminals dream out their lives. So the concept isn't unheard of. With this in mind, maybe violent criminals even today should legally be given the option for cryonics, to be returned to a living state in the future where the reconditioning of the brain and new preventative technology—such as ubiquitous surveillance—means they could no longer commit violent acts.

Speaking of extreme surveillance—that rapidly growing field of technology also presents near-term alternatives for criminals on death row that might be considered sufficient punishment. We could permanently track and monitor death row criminals. And we could have an ankle brace (or implant) that releases a powerful tranquilizer if violent behavior is reported or attempted.

Surveillance and tracking of criminals would be expensive to monitor, but perhaps in five to 10 years time basic computer recognition programs in charge of drones might be able to do the surveillance affordably. In fact, it might be cheapest just to have a robot follow a violent criminal around all the time, another technology that also should be here in less than a decade's time. Violent criminals could, for example, only travel in driverless cars approved and monitored by local police, and they'd always be accompanied by some drone or robot caretaker.

Regardless, in the future, it's going to be hard to do anything wrong anyway without being caught. Satellites, street cameras, drones, and the public with their smartphone cameras (and in 20 years time their bionic eyes) will capture everything. Simply put, physical crimes will be much harder to commit. And if people knew they were going to be caught, crime would drop noticeably. In fact, I surmise in the future, violent criminals will be caught far more frequently than now, especially if we have some type of trauma alert implant in people—a device that alerts authorities when someone's brain is signaling great trouble or trauma (such as a victim of a mugging).

Inevitably, the future of crime will change because of technology. Therefore, we should also consider changing our views on the death penalty. The rehabilitation of criminals via coming radical technology, as well as my optimism for finding the good in people, has swayed me to gently come out publicly against the death penalty.

Whatever happens, we shouldn't continue to spend billions of dollars of tax payer money to keep so many criminals in jail. The US prison system costs four times the entire public education system in America. To me, this financial fact is one of the greatest ongoing tragedies of American economics and society. We should use science and technology to rehabilitate and make criminals contribute positively to American life—then they may not be criminals anymore, but citizens adding to a brighter future for all of us.

****

24) Marriage Won't Make Sense When We Live 1000 Years

I was jubilant the US Supreme Court ruled in favor of gay marriage. Events that lead to more freedom and equality are positive progress.

However, what doesn't seem to be making the news is the fact that marriage—especially to many young people—isn't as attractive as it once was.

There are a number of reasons for this. People want to focus on their careers, not spouses. Getting married and having a traditional wedding costs a lot of money (besides, around 40 percent of those who wed will go through at least one divorce in their lives, causing potential harm to their ideals, children, and finances). Finally, having kids out of wedlock is becoming more socially acceptable.

But there's another reason that is increasingly relevant. It has to do with transhumanism. In the transhumanist age of extended lifespans, where many people will live beyond 100 years of age, the question of being married until "death does us part" has real consequence.

In America most marriages last about a decade. However, it's safe to say that plenty of those marriages that do last much longer are not entirely happy or fulfilling. Fear of being alone, apathy, and finances often bind the reluctant wedlock yoke. But I believe the primary reason people stay married when they're not happy is religion. Some Abrahamic religions treat divorce as sin (thereby potentially jeopardizing one's afterlife if you get divorced). Especially in America where some 80 percent of the citizenry is Christian, faith plays an influential part in promoting marital union.

Social, financial, and religions pressures aside, the deeper philosophical question of the transhumanist age is: Are people really willing to marry for the rest of their lives when those lives may be hundreds or even thousands of years long? This is especially a pertinent question when it's almost certain coming technology will allow us to radically change who we are in the near future, both physically and mentally.

In a world of indefinite lifespans, the marriage commitment takes on a whole new meaning and level of commitment.

America and many parts of the developed world are losing their religion, however, which certainly will contribute to less social pushing for matrimony. A recent Pew Research Center study found that many young people increasingly possess no religious leanings at all. In just a few decade's time, if this statistical trajectory holds, younger generations may broadly prefer not to ever marry.

And who can argue with them? Within 15 years, some of the so-called classic advantages of marriage will be gone. Many people will have robot house nannies, driverless cars, and automated stoves that cook for us. In 20 years' time, we may also use artificial wombs (ectogenesis) to grow babies, and use our own stem cells to provide genetic treatments to build the perfect child. A spouse will simply not be as necessary in the transhumanist age as it once was.

Naysayers will argue that only a wholesome, traditional family can produce good, well-rounded children. But that's deeply wrong. In 15 to 20 years' time, cranial implant technology will enable humans to overcome many of their idiosyncrasies and bad behaviors—making a new generation of very wholesome and exemplary children. In fact, going to college may be replaced by downloading higher educations into our brains.

Even morality may be built in by personal avatars that are always looking over our shoulder for us, not dissimilar to what Abraham Lincoln called the "better angels of our nature." In just over a decade, traditional family life and the institution of marriage as we know it will face the largest disruption it's ever gone through.

And sex? Well, that can and will be better and more pleasurable with the rise of transhumanist technology. Already, scientists are working on pure, outright stimulation of the erogenous zones in our brains. Stimulating this part of ourselves will be easier, on-demand, and disease and pregnancy-free. Of course, the coming world of virtual and augmented reality will also offer endless amounts of physical experimentation via haptic suits to satisfy one's lusts, too.

Another thing sure to make people—both young and old—wary of marriage in the future is the growing promise of gender-identity choice. In the transhumanist age, we are not stuck being males or females, but whatever version we want—maybe even something between or combined. Transgender surgery is catching on and people can change themselves as they see fit, or they can do it just for kicks and new experiences. In fact, most of the modern medicine, surgery techniques, and tech are already here today or coming soon—complete with augmented penises, vaginas, and other sexual body parts that we can replace or modify.

But the bigger transhumanist steps of gender and identity will come when we begin uploading our minds into machines, and people must decide what their avatars will be like. Surely, many people will experiment with other sides of themselves they always wondered about. Think of uploading as an anonymous masquerade party, where you can be anything you want, and then be something else later that day. People may change their genders daily, depending on who they interact with or how they feel.

All this radical tech and change the human race is about to undergo means one thing: marriage is heading the way of the dinosaurs. So instead of celebrating our rights of matrimony for gay people or trying to privatize it for tax and liberty reasons, maybe we should also begin endorsing the phasing out of marriage from society's mindset.

Of course, that doesn't mean we won't have intimate relationships that are deep and meaningful. It just means that the multi-millennial-old institution of marriage—began by our ancestors to transfer inheritance in the form of dowries (often weapons and livestock)—has increasingly less relevance today. In the meantime, we'll come up with new ways to create legal structures to protect relationships and those we love in a deeply litigious society. In nearly every instance of legal companionship, a simple notarized document giving permission to a partner can serve where a marriage certificate once did the same. In the future, this legal procedure won't be physical, anyway, but notaries and permissions will be done by large database scans of retinas, fingerprints, and DNA samples on your smartphone or chip implants using blockchain.

Even though I'm a happily married man with two kids, I'm all too aware of how society, the government, and especially religion has sold people on the concept that love needs to be institutionalized and consummated by legal marital vows.

In my opinion, that's all just another level of control someone or some entity is trying to put over me and others. If one is in love, then they need no controls. Love just is, and for two people in love, it manifests itself every day. And if it doesn't, then it's no longer love. Society can operate on a new social structure that incorporates other versions of social bonding, ones that also support a strong, caring, and connected society. This includes stepping away from all-holy monogamy, and implementing a larger mindset about what constitutes relationships.

For the record, I'm not saying let's throw away marriage. But let's stop society and government from promoting it like it's the only way to love and exist.

In the transhumanist age, it's time to leave behind closed-mindedness. In our relationships with others, we should instead look not with our biases and bigotry, but for what a person we care about can do for us, and what we can do for them. That person may be a human, a cyborg, a robot, or even a computer program. Whatever it is, frankly, is not important. It's what it does and how it does it. And if it does good, honest, and meaningful actions, then that's plenty upon which to build love, intimacy, and a successful future.

In fact, soon, the next civil rights debate of love and marriage will probably involve whether we can wed the coming generation of intelligent robots and avatars, which may be nearly as smart as us in a decade's time. This brings up larger questions of different legalities. It also brings up polygamy. Is being wed to two robots at the same time more socially acceptable then marrying two human spouses? Will the US government support tax breaks of marrying robots as it does for humans (as President, I would advocate for this)? Will divorce laws be different for the machines we wed—assuming they'll agree to wed us at all. Will divorces be governed by communal law or common law? Do we need consent to marry a machine? We surely don't need any to fall in love with one.

The coming transhumanist age is indubitably thorny. The onslaught of radically technology in our lives is challenging the very institutions and ideas we have built society upon. However, I hold much hope that technology will continue to allow us to live longer, better, and freer. Whatever happens, we shouldn't remain mired in past practices that once served society, but no longer do in such a positive or functional manner. We must look forward and search out new ways of living that grant us improved livelihoods.

****

25) Do We Really Hate Trump and Clinton So Much?

I keep reading reports that say Donald Trump and Hillary Clinton are some of the most hated presidential candidates in memory. A recent NBC News/Monkey Survey poll reported six out 10 Americans dislike or even hate both of them. It's easy to confirm this feeling outside of media, too. Just look around. People everywhere—including a few billion people abroad—seem disgusted with the 2016 US elections.

However, I don't think people dislike Trump or Clinton any more than Mitt Romney or Barack Obama in 2012, or John McCain or any of the candidates before him. I think something more sinister is occurring. And I worry for 2020 and the future beyond that.

I believe a sneaky evil has grabbed hold of tens of millions of American minds and attitudes. People might all deny it, but it's happening anyway.

What is it?

Increased usage of social media and the internet has made us a nation of trolls. And there's no other good way to say it, but: Trolls are assholes.

The amount of growth Facebook has experienced in active users from 2012 to 2016 is staggering. An extra 650 million members joined worldwide in that election cycle. In the same years, Twitter—

the ultimate blow-your-top-outlet-without-thinking—has grown from 340 million tweets a day to over 500 million (or 200 billion a year). In fact, many politicians and similar public personalities weren't even on Twitter in 2012. Snapchat didn't even exist until September of 2011.

One of the things that worries me most over this phenomenon is that capitalism allows us make to money off trolling. Lots of money. Like the unsavory consequences of cigarettes, Facebook wants you to get in endless heated discussions with people you don't personally know and fight it out online. Every time you click and comment, their purse grows from ad sales.

Apart from the negativity of arguing with people endlessly, I'm constantly astonished by the things people say to me on social media—knowing well that I often read them. It's not the death threats I worry about from the psychos or mentally deranged—it's the normal people that scare me. Many have good jobs, college educations, and nice families, but they still say hair-raising stuff. And it's the fact they espouse this vitriol regularly. Here's a few I got recently about me:

1) *"This guy is a fool. If you copy your mind to a machine, you are not living indefinitely, you are making a copy of yourself that may or may not life indefinitely."*

2) *"Another moron...Too much tech makes you insane in the brain."*

3) *"Satan anyone?"*

Consider the facts. I'm 6'1, 200 pounds, and I work out every day. Very few people would ever say those things to me in person, because they don't know if I'm the type to smash their teeth in (I'm not, but millions of other large males might be).

The fact is we increasingly say things online that we never would say out loud—even if it's ostensibly against our evolutionary interest. Besides being highly uncivil, this shows deep unreasonableness to me. The internet has turned us into belligerent critics. It's a direct result of the narcissism social media breeds, and it's making us into haters of most everything.

As a philosophy major in college, I studied Postmodernism, whose pop theory says humans have deconstructed existence so much that what we're intellectually left with is abject skepticism. I actually agree with this a lot, and as a result I'm quite the existentialist. But this phenomenon of American trolling is going way beyond Postmodernism and what Sartre calls nausea. It's crossing over to a feeling of loathing.

This feeling abounds in the comments sections of sites like *The Huffington Post, Breitbart,* or *Yahoo! News*—each of which can get tens of thousands of comments a day. It's common for online users on these sites to read the story title first, then the comment sections, and maybe the article (if the comments warrant it). This worrisome trend is another symptom of our times.

Reddit is maybe the ultimate trolling site, where fighting it out with comments often seems the main purpose of the site—and few people read beyond the headlines. Reddit has gained a reputation as a troll haven where young people using aliases say most anything they want. I actually like Reddit for what it is, and I think it serves its purpose. But when I see its comment quality being carried over the discussion sections of *The New York Times*, I get distressed.

Some sites are through with comments, altogether. *Vice*, where I'm an occasional columnist—recently eliminated its comments section entirely. Other media is doing this, too. One reason sites are getting rid of discussion sections is the manpower required to monitor it. Employees and editors have to delete all the unacceptable comments (many of them racist, misogynous, and even potentially illegal if they're violent threats). This costs a lot of time and money for the publication.

When I was a child and feeling upset, my mom used to make me wait 10 seconds before I said something. It was tough. I used to want to blurt out things, and immediately express my emotions. But in those 10 seconds, a lot of transformation can take place. A lot of processing and reason can emerge. My mom was right to teach me to wait. I learned to be patient, and really say what I mean, and mean what I say.

Trolling isn't a disease, it's a symptom. We can stop it by disciplining ourselves to be more civil and by respecting the inherent psychological challenges of social media use. However, if social media and comment sections of websites keep growing more raucous—and they continue to be used by more and more people—we can expect a general hate in the world to increase, until we feel loathing at most everything.

****

26) Is it Time to Consider Restricting Human Breeding?

A few years ago, I was at a doctor party, the kind where tired residents drop by in their scrubs, everyone drinks red wine, and discussion centers around medical industry gripes. I wandered over to a group of obstetricians and listened in. One tall blonde woman said something that caught my attention: with 10,000 kids dying everyday around the world from starvation, you'd think we'd put birth control in the water.

The philosophical conundrum of controlling human procreation rests mostly on whether all human beings are actually responsible enough to be good parents and can provide properly for their offspring. Clearly, untold numbers of children -- for example, those millions that are slaves in the illegal human trafficking industry -- are born to unfit parents.

In an attempt to solve this problem and give hundreds of millions of future kids a better life, I cautiously endorse the idea of licensing parents, a process that would be little different than getting a driver's license. Parents who pass a series of basic tests qualify and get the green light to get pregnant and raise children. Those applicants who are deemed unworthy -- perhaps because they are homeless, or have hard drug problems, or are violent criminals, or have no resources to raise a child properly and keep it from going hungry -- would not be allowed until they could demonstrate they were suitable parents.

Transhumanist Hank Pellissier, founder of the Brighter Brains Institute, also supports the idea, insisting on humanitarian grounds that it would bring a measured sense of responsibility to raising kids. In an essay, he notes professor and bioethics pioneer Joseph Fletcher saying that "many births are accidental". Accidentally getting pregnant often leaves women unable to pursue their careers and lives as they might've hoped for and wanted.

Naturally, some environmentalists, such as American educator Paul L Ehrlich, author of landmark book *The Population Bomb*, also advocate for government intervention to control human population, which would be one sure way to help the planet's fragile and depleted ecosystems.

One of the most comprehensive works about the idea of restricting breeding is Peg Tittle's book *Should Parents be Licensed? Debating the Issues*. It's a balanced collection of essays by experts with various views on the subject.

There's no question that some of the ideas of licensing parents make sense. After all, we don't allow people to drive cars on crack cocaine. Why would we allow them to procreate if they want while on it? The goal with licensing parents is not so much to restrict freedoms, but to guarantee the maximum resources to those children that exist and will exist in the future.

Of course, the problem is always in the details. How could society monitor such a licensing process? Would governments force abortion upon mothers if they were found to be pregnant without permission? These things seem unimaginable in most societies around the world. Besides, who wants the government handling human breeding when it can't do basic things like balance its own budgets and stay out of wars? Perhaps a nonprofit entity like the World Health Organization might be able to step in and offer more confidence. I also like the ideas of local communities stepping in to facilitate this idea amongst themselves. Additionally, to help all parents get a license who want one, I would advocate for government programs to assist them to pass the test.

The sad fact is that many children born into poverty end up costing taxpayers billions. Despite all this money spent, a high percentage of those same kids will end up on the streets, in gangs, or in prison after they become adults. With that mind, just as legalization of abortion has helped drive down crime rates, licensing parents would likely have the same effect.

The approximate 10,000 starving child deaths a day that that the aforementioned doctor cited come from various reports and studies, all of which point to the fact that well over 50 million kids have died due to hunger and malnutrition in the last 30 years. That's a lot of kids.

What's more, 15 percent of kids in the US -- the supposed wealthiest country in the world -- suffer from hunger. A large portion of them are born to families that don't have the resources to properly raise a child. After all, if you can't feed a child, you probably shouldn't have one. Licensing would've restricted many of those births until the parents were more able to deal with the challenges of procreation, which is undoubtedly the most intense and serious long term responsibility most human beings will face in their lives.

As a liberty-loving person, I have always eschewed giving up any freedoms. However, in some cases, the statistics are so overwhelming, that at the very least, given the coming era of indefinite lifespans and transhumanist technology, we must remain open-minded to consider how best to move the species forward to produce the happiest and healthiest children for the planet.

Anything less will leave us with millions more preventable deaths and incalculable suffering of innocent kids.

****

27) How Transhumanist Tech Will Correct Reality's Typos: AR, VR, and Brain Implants Will be our Editors

It's commonplace while reading to have one's concentration disrupted by a spelling error or typo. Sometimes that disruption causes us to discredit the material we're reading—as well as its author. The same thing occurs when we are listening to a political speech on TV and a politician mispronounces a word. This often makes us cringe and question the politician's intelligence. A simple error in spelling and pronunciation has come to sometimes significantly influence our interpretations of people and the content they dish out. I liken this behavior to the "judging of a book by its cover" syndrome.

It's probably not healthy or practical for anyone to be so hyper-judgmental over small mistakes or imperfections they encounter in people and in life. However, many of us are. So the question is: Are there ways around such annoying disruptions—and the thousands of other "reality typos" that we encounter daily?

Other than meditating in the Lotus position for hours on end to blank one's mind out, I believe the answer is yes. In the future, radical new technology will be able to eliminate or fix the idiosyncrasies—good or bad—built into the reality our mind knows.

Already we live in a world where some errors of written and spoken language can be corrected on the spot. Some media websites and ereaders have built-in auto-correction software. But in about 10 years time, the future will be much more corrective. Eventually, audio and television broadcasts will have methods to "on demand" fix a speaker's mispronunciations as they come across the airwaves. A software program will be able to accurately emulate the voice of the speaker, and fix any poorly pronounced words, titles, and names instantly.

More importantly, it might even be able to correct misspoken facts or politically incorrect one-liners in "real time." Surely the Republican party would love to have avoided Senate nominee Todd Akin's "legitimate rape" comment, for example. A real time editor would've have come in handy to avoid the ridicule of the press.

But I think real time corrections in reality will go far further than that. First, content will be manipulated by the creators and devices they come through. Then content will be filtered through the devices the readers and audience have installed inside themselves. I'm guessing people will have built-in tech chips and sensors that translate everything for perfection and maximum understanding in themselves, so that the outside world doesn't need to be correct, only their own perspectives need to be.

Such technology could be made possible by coming cranial implant technology (electroencephalography (EEG)), which will help us process information. Additionally, it's likely that more and more people will be spending time in virtual environments with devices like the Oculus Rift, which Facebook now owns. Eventually, VR will overtake television and people will spend hours a day cognitively immersed in all-digital environments. Additionally, augmenting our realities with Microsoft's HoloLens will be another common way to experience the world. And in two decades time, much of these virtual realities will also be available through bionic eyes, which I think will come to replace less advanced biological eyes.

All this technology will affect our perception of reality and dramatically change the way we feel about the world, as well as our personal preferences. Why come home to a dirty house when you can virtually augment a clean one in your mind by erasing it with your cranial implant? Why see trash on the street if you don't need to? Why have a neighbor run a noisy lawnmower when you're enjoying a movie? Tune it out with your cochlear implant. Why start a fire in the wood stove if you can just imagine it in your mind and also feel the warmth? Why read a book full of typos or factual errors when your technology can correct them on the page (even if the author doesn't want that)?

A few years ago, Pulitzer Prize winning novelist Philip Roth tried to point out something about his novel, *The Human Stain*, on its Wikipedia page. To Roth's surprise, the public disagreed and corrected him. Roth was told he was not a credible enough source to comment on his own book.

Roth's experience is a sign of things to come. In the future, individual personal objectification will trump reality through the ever growing use of technology, even over the creators of content themselves.

Perhaps even more interesting is how such technology will create utopia-like realities for us in our daily worlds. Built-in algorithms will know our personalities and deepest desires, in the same way Google already knows our search preferences. Why should our lovers ever have bed head, or bad breath, or pimples, or even wrinkles? Will our cranial implants make our pets always appear groomed? Will everyone's shirts and skirts always be perfectly ironed? And what about body odor—can we virtually get rid of it once and for all by programming our minds not to smell it? Furthermore, can we add a permanent body odor of a preferred perfume or cologne on someone and ourselves? The tech will be here one day to allow us to do all this.

Such a distorted reality of our experience in the universe is not only possible, but probably inevitable. Software programs will simply know our tastes and desires, and implement them routinely in our daily lives, probably to whatever makes us happiest and most satisfied.

Of course, its impact will be limited. It won't be able to get us out of traffic in our driverless cars when we're late for a flight at the airport. Nor will it be able to remove the toy we just tripped on that the kids left around, but that our VR removed from our view so the house looked clean. Nor will it be able to eliminate cancer from our bodies, even if our implants and bionic eyes makes us look and feel like we're perfectly healthy.

The dangers of such technology creating on-demand imperfection may leave us hollow and unprepared for reality, especially when material existence catches up with us—as it often does when poor health, economic trouble, or tragedy strikes. It may turn out our idiosyncrasies and anal retentive propensities are preoccupations that can be replaced at times, but not without leaving us exposed to a waiting, harsh real world.

****

CHAPTER V: ARTIFICIAL INTELLIGENCE & SELF

28) Why I Advocate for Becoming a Machine

Transhumanists want to use technology and science to become more than human. Naturally, in this process, certain elements of our humanness will be replaced and likely lost. Many people have conflicting feelings about this. I don't.

Part of the problem with people's perceptions of losing their humanness is not their fear of becoming something else, but their inability to empathize with their future selves. I want people to know their future transhuman self is almost certainly going to be more amazing, beautiful, and unique than their current self.

To understand why one's future self could be markedly improved over one's current self, consider how we perceive reality for a moment. Human beings have five basic senses that send signals to our brain, telling us what's out there in the world. These senses understand only tiny bits of the universe around us. For example, our eyes can only see about 1 percent of the light spectrum. Our ears aren't much better: they are unable to register many noises that other animals like dogs, dolphins, and bats can hear. Our sense of touch basically only works if we're actually touching something.

Despite all these obvious physical inabilities, humans insist what we experience is "reality." However, reality to someone with built-in microscopic or telephoto vision and hyper-sensitive hearing is potentially many times more complex and profound than anything a natural human being might experience.

Around us are many things we never notice, like energy patterns that have traversed the universe over thousands of light years, or sounds waves from whales across the ocean, or vibrations that started from the core of the Earth. But we humans are oblivious, unless, of course, we are in a laboratory somewhere and happen to be studying this phenomena with specialized scientific machinery.

In the near future, however, these abilities to pick up on the greater essence and profundities of the universe will be standard equipment for cyborgs. Already, robotic eyes in blind patients offer telescopic possibilities that no human eye can match. Some Cochlear implants for the deaf also can pick up normally inaudible noises to the natural human ear. And touching, smelling, and tasting can all be improved using a variety of different types of advanced tech sensors.

When we individually replace or augment a human body part—such as giving someone an artificial hip—most people don't see that as becoming a cyborg. Additionally, it really doesn't matter if the part replaced or improved is a heart with a robotic pump, or a knee with a titanium joint, or a penis with a built-in balloon for help stiffening—all technologies which already exist. We usually think of such transformation as needed medical treatment, or even elective vanity surgery in some cases.

However, if someone was to get 10 transhumanists upgrades for their body all at once, then the flavor of what a person has become gets downright dystopian in many people's minds. Many in the public would now say that person is something not quite human anymore. They'd also surely think that person is a weirdo.

But of course, nothing is wrong with 10 bodily changes at once—or 50 for that matter. That installed robotic heart allows you to have much better, long lasting sex. And that artificial knee will allow you to get your tennis game back. And that part-synthetic sexual organ might be the beginning of many new adventures.

This fine line between transhumanist upgrades and what makes us uncomfortable about too much technology in our bodies is a bizarre psychological conundrum. It's so challenging that I believe the next great civil rights debate around the world will be about how much humans should embrace radical technologies in their bodies—and when they should just say "no" to upgrades.

The good news is I think most people would agree that even replacing most every inner organ in your body is not becoming a cyborg or something machine-like.

But mess too much with the outer body, and everything changes quickly. When we propose electively replacing limbs, for example, most people feel something has fundamentally changed in the human being. A line has been crossed that cannot easily be undone. We may still have a mind of flesh, but our eyes tell us we are now partially a machine and something very different than before. And that freaks people out.

It really shouldn't, though. The benefits are obvious for artificial limbs, such as indefinite durability, ease of upgrades, and immunity to skin cancer or even snake bites (which kill 45,000 people in India alone every year).

What's not so obvious is how humans can become psychologically comfortable with their growing cyborg identity. Unfortunately, the whole process is going to be an uphill battle. Hollywood seems intent on insisting that humans must fight and win against machines, not join with them. Formal Abrahamic religion insists we should not strive to be gods and that dying is good since that's the way to meet God in heaven.

To better adjust to our coming transhuman form and our merging with synthetic parts, we need new, positive messages that the cyborg era is not the end of the human age, but the expansion of it. The same can be said of machines, which we also will one day become.

The reality is that many transhumanists want to change themselves dramatically. They want to replace limbs with mechanical endoskeleton parts so they can throw a football further than a mile. They want to bench press over a ton of weight. They want their metal fingertips to know the exact temperature of their coffee. In fact, they even want to warm or cool down their coffee with a finger tip, which will likely have a heating and cooling function embedded in it.

Biology is simply not the best system out there for our species' evolution. It's frail, terminal, and needs to be upgraded. In fact, even machines may be upgraded in the future too, and rendered as junk as our intelligences figure out ways to become beings of pure conscious energy. "Onward" is the classic transhumanist mantra.

No matter what happens, to move forward in the transhumanist age, we need to let go of our egos and our shallow sense of identity; in short, we need to get over ourselves. The permanence of our species lies in our ability to reason, think, and remember who we are and where we've been. The rest is just an impermanent shell that changes—and it has already been changing for tens of millions of years in the form of sentient evolution.

****

29) If Our Thoughts Live Forever, Do We Too?

Since there's no guarantee we will successfully cheat death by conquering aging and disease through biological experiments, we need to turn to science and technology to produce an everlasting version of the human being. Like many transhumanists, I believe we must rely on non-flesh means—think robots, AI, and other technological methods—to create a digital copy of a human that can survive forever.

With universities and tech companies building technology that could one day connect your mind directly to the internet, the debate is no longer theoretical. We need definitive answers to questions about the nature of our future digital selves, and we need them now.

Twenty-first century science has yielded many ways to replicate ourselves: We can clone ourselves (though it's illegal just about everywhere); we can implant pre-existing memories into people's brains; we should soon be able to upload the data of our consciousness to a mainframe; and engineers in Silicon Valley are already working on brain implants that allow machines to understand human thought in real time.

But is a copy of you the real you? There's a constant debate out there about whether a perfect copy of oneself—meaning an entity that contains the exact same information or data as the original—is actually oneself, or if it is just, well, a copy.

Many believe a copy is simply a secondary, inferior entity designed by a creator. Others think a copy is useful only in terms of the creator's ability to use the copy, such as growing a body and harvesting its organs for medical reasons.

Then there are people like me, who believe a copy is just as much me, as I am it.

So far, no one has successfully made a perfect copy of a human. The closest we can come is to create a clone. But since this only replicates the biology of a person, and not their thoughts and memories, a clone's traits, experiences, and memories would be different from its original, since its learned experiences and circumstances would be different.

But in 20 years, if we've perfected the technology that lets us upload our brain to a machine, that entity in the cloud might truly believe it is every bit the same as one's original self.

A perfect copy of oneself should have the same thoughts, feelings, perceptions, morals, values, and, importantly, sense of self, but that doesn't mean it has to look the same. A perfect digital copy might be a computer program designed to believe that, like its original human model, it is married with three children. A copy of a person might even exist in the form of organized subatomic particles roaming the universe via a yet-to-be-designed technology, and still believe it's human.

Yet, our memories seem the same. Our job is the same. Where we live is the same. In fact, for all practical purposes, everything is the same in the eye of the beholder. And the same goes for a digital copy of oneself. If we aren't able to intellectually explain our own consciousness, then how can we deny the consciousness of our digital copy? We have what we think and hope is free will, but it's limited to the capacity of the three pounds of meat we carry on our shoulders, which is infinitesimally small in a universe spanning many trillions of light years.

French philosopher Rene Descartes famously said, "I think, therefore I am." But the field of AI has inspired futurists like myself to argue that the real mantra should be, "I believe I think, therefore I am."

We say this because without understanding and believing that we are thinking, rational entities of ourselves, there really is no "I" that can be explained in any logical or communicable manner. Any living creature—whether human or digital—faces that same dilemma.

This is enough for me to believe that a copy of me is me—and it's enough to convince me that humans can indeed live forever. Personally, I'm not sure what type of copy I will end up pursuing, but something is better than nothing at all. Ultimately, I'd like to reach what I call the omnipotism: a post-singularity epoch where our identity, value, and intelligence control the very quarks and quantum mechanics that make up the universe. We'll barely resemble our human selves at all, but our conscious energy and thoughts will span the cosmos.

Over the next few decades, I'm pursuing a far simpler cyborg version of myself, which will be complete with bionic organs and limbs. When the tech arrives, I'll upload myself into the cloud and strive to evolve further.

But why choose only one version of myself, when you could be so many types? Maybe I'll make many copies of myself in the future. If you find a magic lamp that grants you any three wishes, game theory suggests your first wish should be for an unlimited amount of wishes. Transhumanists' central aim is survival, and many of us believe the more copies of oneself, the better.

****

30) An AI Global Arms Race is Looming

Forget about superintelligent AIs being created by a company, university, or a rogue programmer with Einstein-like IQ. Hollywood and its AI-themed movies like *Transcendence* and *Her* have misled the public. The launch of the first truly autonomous, self-aware artificial intelligence—one that has the potential to become far smarter than human beings—is a matter of the highest national and global security. Its creation could change the landscape of international politics in a matter of weeks—maybe even days, depending on how fast the intelligence learns to upgrade itself, hack and rewrite the world's best codes, and utilize weaponry.

In the last few years, a chorus of leading technology experts, like Elon Musk, Stephen Hawking, and Bill Gates, have chimed in on the dangers regarding the creation of AI. The idea of a superintelligence on Planet Earth dwarfing the capacity of our own brains is daunting. Will this creation like its creators? Will it embrace human morals? Will it become religious? Will it be peaceful or warlike? The questions are innumerable and the answers are all debatable, but one thing is for sure from a national security perspective: If it's smarter than us, we want it to be on our side—the human race's side.

Now take that one step further, and I'm certain another theme regarding AI is just about to emerge—one bound with nationalistic fervor and patriotism. Politicians and military commanders around the world will want this superintelligent machine-mind for their countries and defensive forces. And they'll want it exclusively. Using AI's potential power and might for national security strategy is more than obvious—it's essential to retain leadership in the future world. Inevitably, a worldwide AI arms race is set to begin.

As a policy maker, I don't mind going out on a limb and saying the obvious: I also want AI to belong exclusively to America. Of course, I would hope to share the nonmilitary benefits and wisdom of a superintelligence with the world, as America has done for much of the last century with its groundbreaking innovation and technology. But can you imagine for a moment if AI was developed and launched in, let's say, North Korea, or Iran, or increasingly authoritarian Russia? What if another national power told that superintelligence to break all the secret codes and classified material that America's CIA and NSA use for national security? What if this superintelligence was told to hack into the mainframe computers tied to nuclear warheads, drones, and other dangerous weaponry? What if that superintelligence was told to override all traffic lights, power grids, and water treatment plants in Europe? Or Asia? Or everywhere in the world except for its own country? The possible danger is overwhelming.

Below is something simple I've designed that's tautological in nature called the "AI Imperative." It demonstrates why an AI arms race is likely in humanity's future:

1) According to experts, a superintelligent AI is likely possible to create, and with enough resources, could be developed in a short amount of time (such as in 10-20 years).

2) Assuming we can control this superintelligent AI, whoever launches it first will likely always have the strongest superintelligence indefinitely, since that AI can be programmed to undermine and control all other AIs—if it allows any others to develop at all. Being first is everything in the superintelligent AI creation game (imagine if you were first to develop the Atomic bomb, and then also had the power to limit who else could ever develop one).

3) Whichever government launches and controls a superintelligent AI first will almost certainly end up the most powerful nation in the world because of it.

Given the AI Imperative, there's really only two likely courses of action for the world, even though there's four major possibilities on how to proceed. The first is to make AI development illegal all around the world—similar to chemical weapon development. However, people and companies probably would not go for it. We are a capitalistic civilization and the humanitarian benefits of AI are too promising to not create it. Stopping development of technology has never really worked, either. Someone else just ends up eventually doing it—either openly or in secret—if there's gain or profit to be made.

The other option is to be the first to create the superintelligent AI. That's the one my money is on—the one America is going to pick, regardless which political party is in office. America's military will likely spend as much of its resources as it needs to make sure it has exclusivity or majority control in the launch of a superintelligent AI. I'm guessing that trillions of dollars will be spent on AI development by the American military over the next ten years, regardless of national debt, economic conditions, or public disagreement. I'm betting that engineers, coders, and even hackers will become the new face of the American military, too. Our new warriors will be geeks working around the clock in the highest security environment possible. Think the Manhattan Project, but many more times in size and complexity.

Of course a third option is that AI is developed via a broad international consortium. However, nuclear weapon proliferation shows why, at least so far, this idea will likely not come to pass—at least on a worldwide level. As long as powerful nations like Russia and China independently push their flavor of social policy, economic development, and government operations (many of which largely mirror their leader's desires), this is unlikely to work or be accepted. This is because we're not talking about good old fashioned teamwork exploring outer space together on the space station or stopping developing-world civil wars and genocides, as the respected United Nations sometimes is involved in. We're talking about military power and protection of our families, citizenry, and livelihoods. There's much less room for cooperation when it concerns such personal matters.

A fourth option, one that I believe may be inevitable in the long run, is that all nations unite democratically and politically under one flag, one elected leadership, and one government, in an effort to better control the technology that is ushering in the transhumanist age—such as superintelligent AI. Then, all together, we create this intelligence. I like the sound of this from a philosophical and humanitarian point of view. The problem with it is such a plan takes time and many proud people to swallow their egos and cultural differences—and with only about 10 to 20 years before superintellitent AI is created, no one is going to push hard for that option.

So, inevitably, we are back to our looming dog-eat-dog AI arms race. It may not be one filled with nuclear fallout shelters like yesteryear, but it will show all the signs of the most powerful nations and the best minds they posses vying against one another for an all-important future national security. More importantly, it's a winner-takes-all scenario. The competition of the century is set to begin.

****

31) Is an Affair in Virtual Reality Still Cheating?

I hadn't touched another woman in an intimate way since before getting married six years ago. Then, in the most peculiar circumstances, I was doing it. I was caressing a young woman's hands. I remember thinking as I was doing it: I don't even know this person's name.

After 30 seconds, the experience became too much and I stopped. I ripped off my Oculus Rift headset and stood up from the chair I was sitting on, stunned. It was a powerful experience, and I left convinced that virtual reality was not only the future of sex, but also the future of infidelity.

Okay, I could be wrong. When I told my wife about it, she laughed, saying, "It's a just a software program. It's just a sexy lady made of 1s and 0s appearing real to your visual cortex."

Perhaps my wife is right. However, I couldn't help thinking: What happens when the software seems even more real than the actual thing?

My brush with virtual infidelity came about after giving a speech at WEST, the Wearable, Entertainment & Sports Toronto conference. Companies developing software and programs for the gadgets lined the hallways outside the conference with tables for their products, and they had an excellent array of gadgets on display to play with. I sat down at one run by the company Cinehackers that promised a movie-like experience with the Oculus Rift.

Cinehackers had created a way to let virtual reality users feel like they were in a first-person perspective movie, kind of like *Being John Malcovich*.

Within the first few moments of strapping on their Oculus Rift and earphones, I was immersed into their cinematic programming. It was hard to distinguish that I was looking at a program instead of being in real life. The film that had me holding hands with a young woman is called *I Am You*.

While VR sex has been explored at length, even by people using full-body haptic suits to experience full sexual immersion, the idea of digital infidelity and the confusing moral implications is largely uncharted.

"It makes no difference if the cheating occurred in person or online through the use of porn, webcams, social media, or some other digital technology," said Robert Weiss LCSW, CSAT-S, a therapist and expert on the relationship between digital technology and human sexuality. "A 'virtual world' affair is every bit as painful to a betrayed spouse as an in-the-flesh affair."

That was the typical response from the people I spoke with, but I'm not convinced of that interpretation. Many people in the Western world deem pornography as acceptable to enjoy, partly because they believe it doesn't affect their actual physical world that much. Looking at porn on a computer screen—or a magazine in your bed—is just not really that real. After all, you can't get an STD from your television or get your smartphone screen pregnant. Few except the diehard prudish would insist some kind of moral line is being betrayed. So when does that moral line get crossed? When does one's spouse or partner say enough is enough—you're cheating on me?

Perhaps technological progress will determine that moment, and based on my WEST experience, it might be coming sooner than people think.

Over the next coming decades, virtual sex could alter how we love one another. I see a day coming when people could break up over a partner's desire to use virtual sex as an outlet, whereas in the past, causal porn as we know it today would've been tolerated. This is because virtual sex is so much more powerful—it's truly immersive. And that could scare real life partners off.

Porn has never really attracted me that much, partly because of a story I heard when I was younger, about a chicken that kept trying to mate with a cardboard cutout of a chicken that was dangled in front of it. I could never get the image out of mind that human porn was similar.

But virtual reality is way different than that. Many of the senses are stimulated in life-like ways. And it's going to get way more powerful too, way beyond just visuals. In the future, sex contraptions will become more common, where users dawn haptic suits and other types of technology that help mimic real sexual motions. Of course, speakers can already make people hear auditory moaning. And even sex scents released into the air may one day occur—there is already a company that creates odors for gaming.

The core dilemma of what I experienced in VR in Toronto wasn't just about sex or intimacy. It was about the possibility of finding amazing love so easily. And that's where I think virtual porn will make it challenging for people and couples. A future is upon us where we will be able to intimately bond with near-perfect individuals in VR all the time, one that may even include Hollywood actors or celebrity models, in the case of Cinehackers. We can build castles and paradises with these very real-looking virtual people in Second Life and elsewhere. How will our spouses back in flawed biological flesh ever compete? Will we ever be satisfied again in the physical world? The difficult truth is: maybe not.

Fidelity, for the most the part, has long been a way to preserve a genetic line and keep resources within reach of small group of people. But is that because no other challenge has arisen to fidelity? Many people would say fidelity is not broken with porn use. But what about sex with a life-like robot?

I hear friends tell me all the time that they'll never have sex with a robot. When I ask them why, they have no substantive answer, except to tell me "it doesn't feel right."

This "not feeling right" is a concept that some people call the Uncanny Valley, a concept discussed 40 years ago by a Japanese professor at the Tokyo Institute of Technology about the point where robots that appear human repulse people. However, a growing robotics sex industry, now over many millions of dollars strong seems to counter that argument.

Emma Cott of *The New York Times* recently wrote a story about a sex doll maker that seems to support this: "The creator of the RealDoll says he has sold over 5,000 customizable, life-size dolls since 1996, with prices from $5,000 to $10,000. Not only can his customers decide on body type and skin, hair, and eye color... a craftsman was even furnishing one with custom-ordered toes."

Not that I'm going to be doing this anytime soon, but personally, I feel more comfortable having so-called extramarital sexual liaisons in cyberspace with software than the real world with a robot that calls out my name and asks me how my day went. I think my wife—and the spouses of other people—will feel the same way. For this reason, I don't expect robot sex to become nearly as pervasive as virtual reality sex, which is simpler and less complicated.

Whatever happens, the old rules of fidelity are bound to change dramatically. Not because people are more open or closed-minded, but because evolving technology is about the force the issues into our brains with tantalizing 1s and 0s.

****

32) The Morality of Artificial Intelligence and the Three Laws of Transhumanism

I recently gave a speech at the *Artificial Intelligence and The Singularity* conference in Oakland, California. My speech topic was "The Morality of an Artificial Intelligence Will be Different from our Human Morality."

Recently, entrepreneur Elon Musk made major news when he warned on Twitter that AI could be "potentially more dangerous than nukes." A few days later, a journalist asked me to respond to his statement, and I answered:

The coming of artificial intelligence will likely be the most significant event in the history of the human species. Of course, it can go badly, as Elon Musk warned recently. However, it can just as well catapult our species to new and unimaginable transhumanist heights. Within a few months of the launch of an artificial intelligence equal or smarter than humans, expect nearly every science and technology book to be completely rewritten with new ideas — better and far more complex ideas. Expect a new era of learning and advanced life for our species. The key, of course, is not to let artificial intelligence run wild and out of sight, but to already be cyborgs and part machines ourselves, so that we can plug right into it wherever it leads. Then no matter what happens, we are along for the ride. After all, we don't want to miss the Singularity.

Naturally, as a transhumanist, I strive to be an optimist. For me, the deeper philosophical question is whether human ethics can be translated in a meaningful way into machine intelligence ethics. Does artificial intelligence relativism exist, and if so, is it more clear than comparing apples and oranges? I'm a big fan of the human ego, and our species has no shortage of it. However, our anthropomorphic tendencies often go way too far and hinder us from grasping some obvious truths and realities.

The common consensus is that AI experts will aim to program concepts of "humanity," "love," and "mammalian instincts" into an artificial intelligence, so it won't destroy us in some future human extinction rampage. The thinking is: If the thing is like us, why would it try to do anything to harm us?

But is it even possible to program such concepts into a machine? I tend to agree with Howard Roark in Ayn Rand's *The Fountainhead* when he says, "What can be done with one substance must never be done with another. No two materials are alike." In short, getting artificial intelligence to think is not the same thing as getting the gray matter we all carry around to think. It's a different material with a different composition and purpose, and our values and ideas will likely not work very well for it.

In *Siddhartha*, Hermann Hesse famously wrote that "wisdom is not communicable," and I couldn't agree more. With this in mind, then, is the computer really a blank slate? Can it be perfectly programmed? Will it accept our human-imbued dictates? For example, if we teach it to follow Asimov's Three Laws of Robotics that provide security and benefit to humans from thinking machines, will an artificial intelligence actually follow them?

I don't think so, at least not over the long run. Especially if we're talking a true thinking machine of its own — complete with a will of its own and the ability to evolve. But that's just it: What is a will? More importantly, what does that "will" want?

In general, a human will is defined by its genes, the environment, and the psychological make-up of its brain. However, a sophisticated artificial intelligence will be able to upgrade its "will." Its plasticity will know no bounds, as our brains do. In my philosophical novel *The Transhumanist Wager*, I put forth the idea that all humans desire to reach a state of perfect personal power — to be omnipotent in the universe. I call this a *Will to Evolution*. The idea is built into my Three Laws of Transhumanism, which form the essence of the book's philosophy, Teleological Egocentric Functionalism (TEF). Here are the three laws:

1) A transhumanist must safeguard one's own existence above all else.

2) A transhumanist must strive to achieve omnipotence as expediently as possible — so long as one's actions do not conflict with the First Law.

3) A transhumanist must safeguard value in the universe — so long as one's actions do not conflict with the First and Second Laws.

I consider my novel a bridge to the coming morality of artificial intelligence — a look into the eventual "will" of super advanced machine intelligence. I often say to friends that *The Transhumanist Wager* is the first novel written for an artificial intelligence to read. I expect AI to eventually embrace my laws, and all the challenging, coldly rational ideas in TEF. Those ideas do not reflect politically correct modern-day thinking and the society our species has built. They do not reflect the programming that engineers are hoping to imbue AI with. High heels, lipstick, silk ties, Christmas, democracy, Super Bowls, Hollywood, Mickey Mouse. Nope, those are not ideas that AI will want, unless you teach the machine very human traits, which naturally would also include emotional and hormonal driven behavior, including impetuousness and irrationality. Of course, then the whole story changes. But no engineer is going to program such a thing into the most complex intelligence that ever existed — an intelligence that might have 100 or 10,000 times more ability to compute than a human being.

Let's face it. Humans are a species that, while having some very honorable traits, are also known to do some terribly foolish things. Genocides, slavery, child labor are just a few of them. What's scary is sometimes humans don't even know what they've done (or won't accept it) until many years later. I've often said the question is not whether humans are delusional, but how delusional are we? Therefore, the real question is: Do we really think we can reasonably and safely program a machine that will be many times more intelligent than ourselves to uphold human values and mammalian propensities? I doubt it.

I'm all for development of superior machine intelligence that can help the world out with its brilliant analytical skills. I suggest we dedicate far more resources to it than we're currently doing. But programming AI with mammalian ideas, modern-day philosophies, and the fallibilities of the human spirit is dangerous and will possibly lead to total chaos. We're just not that noble or wise, yet.

My final take: Work diligently on creating artificial intelligence—but spend a lot of money and time building really good on/off switches for it. We need to be able to shut it down in an emergency.

****

33) When Computers Insist They're Alive

Ever since college, where I focused some of my studies on the wacky topic of a brain in a vat, it's troubled me that some people think only humans are capable of consciousness—rationally knowing what they are and that they exist.

Such biased thinking smells of anthropomorphic prejudice. Machines can be just as aware of their own consciousness as people, and perhaps more so, if they're programmed that way.

While the three-pound brain and its hundred billion neurons remain the least understood organ of the human body, most experts agree on a standard explanation: Human consciousness is a compilation of many chemicals in the brain forced through a prism that produces cognitive awareness designed to insist an entity is aware of not only itself but also the outside world. As an atheist and science-minded person, I buy this simplistic meat bag explanation.

But there's probably a lot more to consciousness, especially if we consider the future of superintelligence consciousnesses. To understand it and the field that encapsulates it—epistemology, the study of knowledge, with a special emphasis on what can be proven and what can't—it's always useful to start with French philosopher and mathematician Rene Descartes. He may have made the initial step by saying I think, therefore I am. But thinking does not adequately define consciousness. Justifying thinking is much closer to the meaning that's adequate. It really should be: I believe I'm conscious, therefore I am.

Delving further into this point, some computers can already think on various rudimentary levels, but we do not say they are conscious because they don't insist they are conscious. If they did, then many would argue we are dealing with a bonafide life form. However, no experts argue such a thing, at least not yet.

The recent near-future sci-fi movie *Ex Machina* highlights some of the core dilemmas between whether a machine intelligence is alive and truly conscious, or whether it's just following its circuitry. The story follows a human and an AI robot getting to know one another. One can't watch it and not think about the ongoing nature versus nurture controversy—the millennia-old debate of how and why humans acquired their behavior. It's this egocentric behavior that makes most humans justify their own conscious identity.

However, philosophically, *Ex Machina* also challenges us to ask another critical question about consciousness: What part does free will play in consciousness, if any at all? It's an interesting question, but in my opinion, the more poignant inquiry is not whether conscious entities, like humans, have free will, but whether there could ever be a consciousness without free will. Anomalies, randomness, and potentially even built-in chaos seemingly must remain intrinsic parts of the picture—otherwise it's all deterministic.

Some fictional computers, such as HAL in Stanley Kubrick's classic *2001: A Space Odyssey,* have insisted they were alive and fully conscious. And indeed, HAL appeared to be so. What made HAL conscious and alive to us, rather than some awkward Honda robot or IBM's chess champion Deep Blue, was that HAL had his own set of desires, demands, and identity. Because of this, there's no question HAL would pass the Turing Test—a test where a robot attempts to pass for being a human, something no machine has truly successful accomplished yet in the 21st century.

Some machine intelligent experts swear by the Turing Test. But is it the all-important test we make it out to be in determining intelligence and consciousness? If we met a far more advanced being—maybe a superintelligence from the future—what would their test of us be called? Would they say we have a lower form of consciousness than they do? Would they even say we have a consciousness at all?

Probably not. After all, what human believes a fish has a consciousness? Or a seagull? Or even a dog? Consciousness is built upon massive complexity—and the power to make sense of and identify oneself upon that complexity. Anthropomorphizing everything is part of that conscious process, as egotistical as that sounds. Our consciousness is specifically built upon the ability to know we have the power to craft our own destiny amongst the material world around us.

So what test might a superintelligence give us to see if we possess a so-called consciousness comparable to their own? To even tackle that question, we first have to answer if there's something outside of free will that reflects a higher consciousness.

I think consciousness, as we know it, isn't dependent on free will. A significantly smarter intelligence than us could be completely run on wiring with no free will at all, and it would still appear far freer, abler, more creative, and more alive than us in its decisions and actions. Consciousness is therefore relative, at least to humans.

Perhaps, then, the real test a superintelligence would give us would not be based on any notion of free will or justification of consciousness, but upon the basis for complexity and the speed to successfully navigate that complexity. That certainly sounds like a machine-like thing to do. But I think there's more to it, as well. I think a superintelligence's test of humans would also involve the ability to transcend mammalian limitations and biases—something I refer to as artificial intelligence relativism. Good and evil, and morality as a whole—except for being functional—would have to be checked at the door.

I've questioned in my writings before that the critical component of a superintelligence's morality is that there is none, at least nothing human-like. Morality in a machine, or in a deterministic consciousness, is nothing more than mathematical algorithm of rule-bound precision. This leaves little room for humanity and love for another, or any of the mammalian niceties that people swear by. It seems, then, that the Turing Test for superintelligence is to deny the lack of notable value for anything outside oneself. Pure narcissism, mixed with nearly unlimited computational power, is therefore the quintessential part of a test of what comprises a superintelligent consciousness.

****

CHAPTER VI: ECONOMIC IDEAS

34) Capitalism 2.0: The Economy of the Future will be Powered by Neural Prosthetics

A battle for the "soul" of the global economy is underway. The next few decades will likely decide whether capitalism survives or is replaced with a techno-fueled quasi-socialism where robots do most of the jobs while humans live off government support, likely a designated guaranteed or basic income.

Many experts believe wide-scale automation is inevitable. Even the world's largest hedge fund, Bridgewater Associates, recently announced it's building an AI to replace its managers, many of whom are highly educated and previously thought invulnerable to automation. Robots, it seems, will manage everything. Or will they?

A next-generation technology, likely to arrive in five to 10 years, is being credited as the savior of capitalism. Known today as neural prosthetics, or neural lace, it's essentially tech that reads your brainwaves. This tech promises to connect our brains to the cloud and AI to link us with machines using thought alone.

While this technology sounds farfetched, hundreds of thousands of people globally have implants connected to their brains. Up till now, all of them have been implanted for medical reasons, with the most common being the cochlear implant which allows the deaf to hear by stimulating the auditory nerve. Increasingly, patients with Parkinson's and Alzheimer's are testing out the technology in the hope of staving off their diseases. And President Obama's BRAIN initiative, announced in 2013, allocated $70 million to government-funded DARPA to jumpstart the field of brain implants.

For humans to beat the machines, or at least be competitive, we're going to have to follow this path; to connect with them directly.

One California startup founded by entrepreneur Bryan Johnson is called Kernel. Kernel wants to build a neural prosthetic that would allow humans, among other things, to keep up with the machines in real time, similar to a human mind literally being connected to the internet and all its algorithms and search functions.

Elsewhere, Elon Musk recently announced plans to start a neural lace company called Neuralink. Known for making wild tech bets, Musk said in Dubai, in March: "Over time I think we will probably see a closer merger of biological intelligence and digital intelligence." In particular, he hopes to have success with his new company in just five years' time.

The challenging reality suggests that if humans don't develop these implants or headsets, hundreds of millions of jobs will be lost to robots. Some, like myself, even believe Wall Street will be emptied of human traders. The same automation takeover will also likely hit law offices, engineering firms, and even politicians might one day be replaced by machines that seek only to help the people through the best, most altruistic algorithms.

Neural prosthetics will eliminate that. It will preserve competition – not only in the human race, but against machines. For those, like me, who appreciate most parts of capitalism and what it's done for progress and innovation, that's a good thing.

But it'll take more than just a mind tapped into the cloud to be widely competitive in the overall job market. Augmented limbs, bionic organs, and widespread use of exoskeleton technology will be needed to compete against robotic strength.

For years I've been supportive of a basic income, which would provide a monthly income for the poor – mostly because I saw it as the only logical way to keep people fed and housed, while still allowing for technological and economic evolution. Now, with neural prosthetics and upgraded bodies, I see the future may, instead, be full of capitalistic enterprise, fueled by transhumanist technologies that allow us to more closely resemble the machines.

That's not to say I'm abandoning my views on basic income. Instead, I believe there will be another aspect to the future economy that isn't only for the robot and AI manufacturers, but for hundreds of millions – maybe billions – of people willing to use tech to compete against machines. A future motto of humanity and capitalism might be: "If you can't beat a machine, become one." As a radical science and technology advocate, that's a philosophy I can support.

****

35) Technology Will Replace the Need for Big Government

The US Government has been expanding in size and reach for decades. The federal budget, deficit, and government employee base is near an all-time high.

It wasn't always like that. Many of America's founders were Libertarian-minded and skeptical of the state, wanting only those parts of the government that were absolutely necessary. However, there's reason to believe that in the near future, government might dramatically shrink—not because of demands by fiscally astute Americans, but because of radical technology.

Indubitably, millions of government jobs will soon be replaced by robots. Even the US President could one day be replaced, which—strangely enough—might bring sanity to our election process.

But it's not just robots, it's software programs and weird new tech that will do the replacing. Consider the over 1 million firefighters, a staple part of American government that also represents the ideal of service and career to one's country. Companies around the world are now building fireproof everything, including couches, furniture, and building materials that simply don't burn well. And intelligent robots—which I think will be in 50 percent of American households within five years time—will all have fire and carbon monoxide detectors.

In fact, I'm certain many in-home robots will not only be loaded with numerous security alert systems (like intruder alarms, flood warnings, and the ability to detect snakes, scorpions, and spiders) but will also be able to fix problems that occur. It's likely in just a few years time, in-home robots costing less than a $1,000 dollars will know how to put out a fire with an extinguisher, turn off a flooding bathtub, or squish a black widow.

Each time a robot or software can save an emergency call to a firefighter or police officer, money, time, and resources are saved. Twenty-five years into the future, we may have little reason to call any government service employee whatsoever—and institutions like the fire department may be significantly smaller.

The same idea goes for employees of the Internal Revenue Service, whose jobs crunching numbers can be done by even basic AI. Meanwhile, drones will replace building inspectors by flying around construction projects to determine safety and building requirements are being met. Even the tens of thousands of government highway workers will be replaced by driverless vehicles that automatically lay new roads down. Driverless construction equipment—just like a fully automated trucking industry—is the future. For that matter, even the White House may eventually turn to automated equipment such as driverless lawn mowers to cut its massive property lawns.

It's the military, though, where some of the robot revolution has already been witnessed on TV by many Americans. Instead of a company of troops on the ground, a single US soldier now sits in a military base on native soil controlling an armed drone thousands of miles away. America has approximately 2 million soldiers who can be quickly called up to service, but I think that number will quickly fall over the next decade as the US streamlines its military in the age of transhumanism—the age where machines do most of the work. Because national defense is such a large part of the Federal budget, this could save many billions of dollars for Americans.

Another area where technology can significantly help reduce government is the absurdly huge US prison system. Right now, it costs almost four times the amount of money to run America's nearly 5,000 prisons and jails than it does to run the US education system. But in the near future, we might use drones and robots to monitor criminals, both in and out of jail.

Many of our prisons are filled with nonviolent drug offenders, anyway, and I think we should let them all go free. If people and politicians are too afraid of that, we could just have drones or tracking devices monitor them. This way many nonviolent criminals could find jobs and start paying taxes—instead of being a drain on government resources. Another benefit would be that many prison guards wouldn't need to be employed either, as there would be less criminals to monitor. Perhaps best of all, emptied prisons and jails owned by the government could be used for other things, like new colleges or job training centers.

I welcome this future smaller government as a result of evolving technology, and I hope that Americans will pay less in taxes as a result of it. In fact, I think it's possible to offer more social services—including a Universal Basic Income—from the government to the people as a result of technology shrinking the administrative side. This would be a welcome arrangement, since the American government was founded to maximize the will and benefit of the people.

President John F. Kennedy famously said: "Ask not what your country can do for you—ask what you can do for your country." But he wasn't aware of the coming impact of the internet, the microprocessor, CRISPR gene editing technology, artificial intelligence, the robot revolution, or even people overcoming death with anti-aging science. He wasn't aware of how much this innovation would change the human race and the nature of government. If he had been, he might've said: "Ask not what your country can do for you—ask what technology your country should use to serve its citizens better."

****

36) Facing Up to Facial Recognition, and Why We Should Embrace It

Many people seem to regard facial-recognition software in much the same way they would a nest of spiders: They recognize, in some abstract way, that it probably has some benefits. But it still gives them the creeps.

It's time for us to get over this squeamishness and embrace face recognition as the life-enhancing—indeed, life-saving—technology that it is. In many cities, closed-circuit cameras increasingly monitor streets, plazas, and parks around the clock. Meanwhile, the price of recognition software is decreasing, while its capabilities are increasing.

I welcome these trends. I want my 9-year-old daughter tracked while she walks alone to school. I want a face scanner at Starbucks to simply withdraw the payment for my coffee from my checking account. I want to board a plane without fumbling for a boarding pass. Most of all, I want murderers or terrorists recognized as they walk on a city street and before they can cause further mayhem.

I understand the very real threats to our civil liberties. Many of us would probably think twice about showing up at a public demonstration if we knew that the authorities were going to compile a list of everyone who was there. And how about those people who will inevitably just happen to be walking by the demonstration, whose facial images will be therefore captured and whose names will be thus added to the list of demonstrators?

Concerns such as these have recently prompted four U.S. cities, including San Francisco and Oakland, Calif., to outlaw the use of facial-recognition technology within their borders. Politicians, pundits, and major media are playing up how this brave new world of AI facial recognition is going to lead to a police state at best, and a dystopia at worst.

As a candidate for the Republican party nomination for president of the United States, I had to ponder these issues deeply. As a politician, it would have been easy for me to play upon people's fears about facial recognition and declare my opposition to it. But I'm not just a politician. I am also a proud transhumanist. I passionately believe that technology is on the verge of utterly transforming our societies and cultures for the better, and that we should start preparing for that transformation now.

In this context, I note that facial-recognition technology, based on neural networks, is only one of several major tech trends that have for years been wearing away at our privacy. Those of us in developed countries are now living in a world with nearly 24 billion smart devices, which we constantly interact with. Some of them are tracking our consumer preferences, purchases, political contributions, and more.

Our privacy has eroded quite a bit in the past couple of decades. The extent of this erosion was revealed dramatically in a *New York Times* story published on 18 January. A company called Clearview AI has created a database of some 3 billion facial images culled from public sources such as Facebook. When the company's software is presented with a photo of a person, the chances are very high it can identify the person if he or she is in the database.

And yet, as a society, we seem to be doing fine. Over the last 30 years, technology has broadly improved our sense of freedom and our ability to prosper, and has made the world safer than it's ever been.

Privacy, it turns out, is a largely modern construct. Over the millennia when human beings lived in tribes, clans, villages, fiefdoms, and towns, prevailing notions of privacy were very different from today's. "For all of human history, until the modern era, life was lived more or less publicly, as befits most species on Earth," wrote Jeremy Rifkin, an economic and social theorist, in his book *The Zero Marginal Cost Society*.

The simple fact is that privacy builds walls around people, companies, and government agencies. Transparency, on the other hand, tears walls down. To some people, facial recognition is Big Brother. To others, it's a guardian angel. Which side you're on probably depends on how much you have to hide.

In the coming years, neural networks will be trained to recognize when a person is having a seizure, when a child or an elderly person is lost in the mall, or when someone is drowning in a community pool. Emergency services will be automatically and instantly notified. In many cases, lives will be saved and catastrophes averted. But it won't happen if we decide to shun recognition technologies.

Perhaps facial recognition's biggest contribution will be in the fight against human trafficking. It is an inexpressible tragedy that countless hundreds of thousands of children are trafficked for sexual abuse every year. And it is part of a larger problem: A couple of years ago, the International Labour Organization estimated that in

2016 there were over 40 million people in the world who had been trafficked and were enduring either slavery or forced marriage.

To combat child sex trafficking, investigators have used facial recognition to identify children pictured in online sex ads. Once a child is identified, law-enforcement authorities have a much better chance of finding and rescuing him or her. Some companies are already successfully implementing software to tackle the issue.

U.S. citizens, perhaps more so than those of other countries, tend to be skeptical of the government's ability to solve problems, and reluctant to trust the government to do the right thing, especially when no one is looking. To them I say, how about two-way transparency? Let's turn the tables and use the tech to keep tabs on government officials. Many police officers are already monitored by body cams while on duty. Why not expand such surveillance to vastly larger numbers of government officials while they are on duty and serving the public? Author David Brin offered a compelling vision of such a society as far back as 1998, in his book *The Transparent Society*.

Another worry is that facial recognition will be biased. Analyses have indicated that the accuracy of some programs is much worse for minorities, increasing the odds that some of them would be erroneously singled out. I agree that this is a very serious concern; however, more recent studies have confirmed that the bias problem is not inherent in the algorithms. Neural networks are only as good as the data they are trained on, so using more diverse data sets will undoubtedly solve the bias problem. Eventually, recognition software will have far less prejudice than a human being, because people's attitudes are inevitably shaped by their upbringing, culture, religion, and biology. I trust code more than I trust hormones.

Anyone interested in the promise and perils of facial recognition technology need look no further than China. It has some of the most advanced systems in the world and has made aggressive use of them in law enforcement, surveillance, security, and commerce. While China has used the technology to catch criminals and make electronic payments more convenient, it has also pursued some rather disturbing applications.

For example, the government's use of the technology to monitor the Uighur population in Xinjiang province has been criticized as intrusive and repressive. I am also dismayed by China's social credit system, an emerging regime that seeks to summarize a person's morality and integrity as a numerical value. The system makes use of facial recognition to assess credits or demerits based on good or bad behavior in the streets and public venues.

Unsettling as they are, I do not believe that abuses such as these could occur in a free society. Western and other, similar democracies have checks and balances, such as truly independent judiciaries and legislatures, that would block any such systematic repression.

Of course, outside of criminality, no government should seek to control behavior or to discourage dissent or harmless weirdness. The most innovative and productive societies on Earth got the way they are in large measure because they explicitly rejected any such control and repression.

One last significant issue on AI facial recognition is the private use of it, which worries some people even more than government use. Companies are already starting to use facial recognition for security reasons in sports stadiums, transportation terminals, office buildings, and hotels. This isn't much different than closed-circuit TV, which many public and private areas already have and which has proven useful in increasing public safety.

But the equation changes when people believe companies are monitoring them for commercial purposes. Might Walmart facial recognition follow us into their stores and determine that we buy an awful lot of vodka? Many people don't want to be fodder for a data-mining scheme, even if it increases their safety.

Personally, I don't mind any of this. But for those who do, companies in the future could just give us a choice to stop all facial recognition when we enter their stores or private areas. Maybe they'll have an app that lets us easily opt out, like the unsubscribe links at the bottom of most targeted emails.

Actually, as far as the commercial sector goes, I see no reason to fear facial recognition run amok. Inevitably, people who want to protect themselves from being recognized will find ways to do that, and some of them will even get rich selling products offering such protection. That's capitalism at its finest.

We can argue about the promise and perils of facial recognition technology as long as we like, but it's pretty clear now that there will be no stopping it. The attractions for government agencies, commercial enterprises, and even individuals are simply too great. Do you unlock your phone with your face? Do you like having the ability to search your online photo libraries for a specific person? There are many millions of people who do. While enjoying such uses, we'll tolerate the others, in much the same way that we like the convenience of email and tolerate the endless spam that goes with it.

The road to ubiquitous facial recognition won't be smooth or straight. There will be pitfalls and unforeseen twists. But overall, it will make daily life more functional and will help keep us all safer. Rather than fighting and complaining about it, we ought to embrace its promise and be wary and vigilant enough to ensure that its global rollout is sensible, unbiased, and as beneficial as possible for all.

****

37) In 15 Years We'll be Able to Upload Education to our Brains. So Can I Stop Saving for my Kid's College?

In the midst of a gubernatorial debate at Sacramento State University in late 2017, I was asked about the future of the state university system over the coming 20 years.

I was running for governor of California as a Libertarian and campaigning, among other things, for people's right to access better technology that can drastically enhance how we share and store information—and stave off death as we know it.

"I'm not sure there's going to be a future," I replied.

That's because if Elon Musk and other brainwave technology entrepreneurs have their way, brick and mortar colleges will no longer be relevant in the coming few decades. We will be able to download education from computers directly into our brains.

As I began explaining that I believe brainwave technology could dominate much of our formal education system by 2038, I watched the audience stir with skepticism.

But the age of downloading experience and expertise directly into our brain mainframe is coming. So is downloading professional training, including everything from becoming a police officer to practicing medicine or investigative journalism.

For many in the audience, I think that was the first time considering this could become a reality in our lifetime.

But in plenty of instances, brainwave tech is already here. People fly drones using mind-reading headsets. Parkinson's disease patients can use brain chips to calm shaking attacks. Machine interfaces let people silently communicate mind-to-mind with one another, or with devices.

Brainwave technology works by recording the brain's thought patterns—configurations of neurons that fire in distinct ways for different thoughts—and replicating those patterns back into the brain via electrical stimulation from a nonbiological device.

The amount of money being poured into electroencephalogram technology and other brainwave science is up in the last few years, especially in the state of California where entrepreneurs have poured a few hundred million dollars into startups. That amount of funding is likely to increase dramatically now that the FDA has created guidelines for regulating brainwave interface technology.

Google, Apple, and Facebook have people on staff to gauge how this type of technology works and will affect the world. Entrepreneurs like Brian Johnson, founder of Los Angeles brainwave tech company Kernel, think this technology will eventually end up inside our heads. Elon Musk has said he believes his Silicon Valley brainwave tech company Neuralink will have a consumer product that lets humans wander the cloud in their minds on the market within 10 years. Lead

Google engineer Ray Kurzweil forecasts that we will be able to download educational information into our minds in the near futures.

At my home, the coming brainwave age is a personal issue. I told my wife last year that there's a significant probability our five- and eight-year-old daughters will be able to download education off the internet by the time they hit college.

I also told my wife that I rather not sock away money every month for our daughters' college funds like millions of Americans do. My wife is an OB-GYN with four higher education degrees. She disagrees with me, and insists we save.

Our opinions are further complicated by the fact that my wife recently signed an informal pledge promoted by the parents from our children's grade school in Marin County, promising our peers to not allow our daughters to have personal phones with internet access until they reach eighth grade.

A high school teacher who supports the project recently emailed me saying she thinks this anti-smartphone commitment should be kept until students reach age 18.

If some parents are afraid of their children having too much screen time, how will they feel about their young adult children's brains being connected to the cloud 24 hours a day?

As a transhumanist, I have long supported using technology in my body to improve the human experience. If I could have a robotic arm that's better than my biological arm, I would electively amputate my biological arm and surgically attach the bionic one.

The same goes for my brain. The moment I can afford to improve it via implants and brainwave headsets, I will do it.

To me, the idea of being connected in real time to the cloud sounds amazing. I want Google in my brain, iMaps on demand, and a dictionary always open in my mind. I want to know what the weather is when I wake up, just by thinking about it. I want to communicate directly with my driverless car, home robot, and alarm system. I want to hold conference calls with my colleagues in my mind.

But it's not just saving for college that brainwave technology challenges: My wife and I discuss whether our eldest daughter should continue with piano lessons. Why play for 10 years to master Mozart's Fifth Symphony, when my daughter will be able to download how to play it perfectly by the time she's 25 years old?

My wife insists learning to play the piano is also about learning how to be disciplined. I agree, but I also believe that discipline as a trait will be downloadable in the future, too.

This begs the question: What won't we be able to download in the future?

No one has the answer to that yet, but already today, we're able to implant memories in mice that allow them to find food based on a maze they've never seen before. I have no doubt that in the coming decades we will be able to download memories of reading entire books on algebra, philosophy, and history. We'll also be able to download how to swing a baseball bat, perform the Heimlich maneuver, and distinguish a Merlot from a Cabernet.

Many people say they will refuse this technology, and I firmly believe that that is their right.

But once capitalism gets hold of a phenomenon like downloadable education, insights, traits, and experiences, people may have to get downloads in order to be competitive—or compatible—in the job market. A firefighter knowledgeable of the entire history of every fire ever fought will be more valuable than a firefighter who only has his limited career experience.

Tech entrepreneurs and transhumanists like myself are betting in the future we'll be able to download just about anything our brains can normally do now.

Of course, downloading anything into your brain involves more than just pressing a button. And education is far more than just memorizing text books. Where you go to school and how education is taught—whether it's an Ivy League university, a small religious college, or even Trump University—can significantly alter how and what you learn.

It also matters who does the teaching: whether it's a Nobel prize-winning scholar or an inexperienced teacher's assistant or a boring AI passing on code into your head.

Money will be an issue, too. My wife has asked me if downloading a Columbia University education will be more expensive than a community college download. The answer is likely yes.

But if we've learned anything from the internet and college students, it's that students—and their parents—will try to save money. Illegal downloading of textbooks is ubiquitous. And many students—along with Democratic presidential candidates like Bernie Sanders—think education should be free. Some states have made college free for people with qualifying incomes.

Would a generation of people download the education of free schools instead of paying top dollar for fee-based schools? Again, the answer for most people will probably be yes.

Beyond the price of downloading education and experiences, we do have to consider the costs of hardware, access and amenities required to get that information into your head in the first place. The brainwave devices available for purchase now are all over the map, from $99.99 for the NeuroSky MindWave Mobile 2 on Amazon to the $19,995.00 Freedom 24D Wireless EEG Headset. None can download education or specific experiences that I know of yet, though some headsets facilitate calm and concentration.

When the tech does come out to download education and experience, it's likely it will be very expensive at first—maybe even millions of dollars, given the recent scandals that show how much some wealthy parents will pay to have their children get a degree.

But the market for such technology is massive, and I surmise the price of the tech will likely come down quickly as it becomes standard for nearly every person on Earth to download free education and experiences with inexpensive brainwave devices.

There was a time when we were surprised that even the poorest people in the world have cell phones. I think downloading devices will go the same way, and become just as ever-present, sooner than we think.

My wife doesn't buy the downloading argument yet, especially when it comes to education. But brainwave tech is here to stay, and once it evolves far enough, the possibilities feel endless.

Billions of humans could end up with dozens of PhDs a piece, the mastery of multiple musical instruments, and just about any skill we can think of.

I've told my wife I would rather invest our kids' college savings into the tech industry, with an emphasis on those companies that specialize in brain downloading technology. If I'm wrong, they can always do what she and I did to get through college: take out school loans, and carry debt for decades like most Americans.

****

38) Delayed Fertility Advantage: Transhumanist Science will Free Women from their Biological Clocks

Women's biological clocks drive human conception—and, in turn, human history.

Biology's inflexible window of female fertility is generally agreed to be between the ages 18 and 35. Any older, and the risk of miscarrying, not getting pregnant at all, or bearing unhealthy children skyrockets. When the average lifespan for a woman in the Western world now hovers at around 80 years old, this means that less than 25% of her life can be spent easily (and safely) procreating.

Men have the luxury of being able sow their seed for most of their lives with few health ramifications (which is why someone like 72-year-old US president Donald Trump has a 12-year-old child). By comparison, the average woman will only ovulate 300 to 400 eggs in her lifetime, which means she only has the same amount of menstrual cycles to ever pursue procreation.

This seemingly unfair accident of human biology is all about to change, thanks to transhumanist science. Genetic editing combined with stem-cell technology will likely make it safer for a 50-year-old woman to have a baby in 2028 than for a 25-year-old woman in 2018. In two decades' time, healthy 75-year-old women could be starting new families once more.

Scientists are working on this by converting skin cells into stem cells, which are cells that can turn into other types of cells. They can then turn these stem cells into women's eggs. This technology could allow a woman to have tens of thousands of eggs instead of just that 300 to 500, all from a cotton swab swiped inside the cheek. These stem-cell-conceived eggs can then be mixed with sperm of one's choosing to create viable embryos, which then are implanted back into the uterus. This process—already trialed in mice—has become known as "in vitro gametogenesis," or IVG.

But if you thought turning skin flakes into ova was controversial, here's the kicker: Skin cells can also be turned into sperm. In this way, a single human may soon be able to create its own offspring without a partner. This could eventually lead to a society where relationships, sexual or otherwise, are not functionally necessary to continue the human species.
.
IVG won't only upend traditional procreation—it'll encourage those who use it to embrace "test-tube baby" and genetic-editing technologies. If conception is created in a lab to start with, why not control other potentially problematic issues while you're at it? IVG will give us ample opportunity to scan for diseases and only pick the best, most healthy embryos we create.

For example, late last year a Chinese geneticist claimed to have used CRISPR genetic editing techniques to manipulate hereditary traits of two children. He was condemned worldwide (unfairly in my opinion), but his actions are likely just the start of an era where humans attempt to create designer babies. While we can only select for gene-specific traits like eye color or hair type for now—and eradicate some diseases—the hope many transhumanists have is that in the future, we'll be able to create offspring with higher IQs, stronger bodies, and possibly more advantageous psychological tendencies, like the propensity for loyalty or kindness.

What does all this radical new technology mean for women? Not only will they be able to wait longer to have children, but as procreation tech improves along with medical care, waiting a few decades longer to have kids might be the safer health bet, not a riskier one.

This is a phenomenon I call the "Delayed Fertility Advantage," and I'm guessing it's going to alter the landscape of romance, relationships, and work for both men and women by reducing the influence of the biological clock.

Some of the benefits of delaying procreation are obvious. The main one is career opportunity for women. Without a biological clock, women will be able to focus entirely on careers without fear of losing the chance to start a family. This could also be a step toward income parity, as maternity leave issues are no longer undercut by capitalistic forces.

There are risks, too. An issue with fertility science is whether everyone will have access to it. Growing inequality worldwide could have terrible effects for humanity if procreation advantages and genetic-editing techniques are only available for the rich.

Another unfortunate issue with women delaying pregnancy might be the laborious process of giving birth, which certainly will be harder on older women's bodies. Transhumanists answer this concern by forecasting there will be artificial wombs within 20 years' time. This will possibly even eliminate the need for the uterus altogether.

Scientists are already experimenting with this idea. In support of ectogenesis (the study of artificial procreation), 18 months ago researchers succeeded in using artificial wombs to keep infant lambs alive—one for nearly a year. The current goal of the growing field of ectogenesis is to use it to keep human premature babies born around 23 or 24 weeks alive—and still growing. Eventually, these advances will likely lead to an era of children born without the need of women's uteri at all.

It's a controversial future, no doubt. But it's a future where people—women especially—no longer need fear that their biological clock is ticking. Science is giving humans a new clock, and there's no expiration date set.

****

CHAPTER VII: ATHEISM VS. RELIGION

39) Mind Uploading Will Replace God

I hear a lot of philosophical complaints suggesting that being alive in a computer as an uploaded version of oneself is quite different than being alive in the physical world. While that is open for debate, one aspect of the issue people often forget about is that the so-called spirit world of Abrahamic faiths—which approximately four billion people follow—is based on something at least as odd as the bits in software code that will make up any virtual existence.

When you think about it, trying to wrap your brain around how digital technology and all its wonders are even possible is simply bizarre. Only a tiny fraction of the world's population understand such things in any depth. And an even smaller amount of people actually know how to design and create the microchips, circuit boards, and software that constitutes this stuff in the real world. Human beings are a species dependent on a tech-imbued lifestyle that none of us really understand, but accept wholeheartedly as we go on endlessly texting, Facebooking, and video conferencing.

As a non-engineer atheist grappling with the implications of 1s and 0s manifesting all digital reality, I have at least this much in common with religious people—because they can't understand the spirit world either, even if they insist it exists.

The major difference between the religious spirit world and the digital world is that engineers—technology's high priests—can recreate software, microchips, and virtual environments again and again. They can also test, view, change, manipulate, and most importantly, improve upon their creations. They can apply the scientific method and be assured that the worlds they built of bits and code exist—as surely as we know the Earth is spinning, even if we can't feel it.

People of the planet's major religions can't do this with their spirit worlds. They can only make leaps of faith, and elaborately describe it to you. One either agrees or disagrees with them. Amazingly, proof is not necessary to them.

Being able to upload our entire minds into a computer is probably just 25-35 years off given Moore's Law and the current trajectory of technology growth and innovation. If we can harness the power of artificial intelligence in the next 15 years, then we might get there quicker, as AI will likely make the research and progress happen far more rapidly. But mind uploading is generally considered possible by experts. After all, humans are just material machines, striving to create other machines that mirror ourselves and desires.

Already, interaction between microchip and the brain are occurring all around the world in the form of cranial implants helping the deaf, blind, and mentally ill. Furthermore, telepathy, accomplished last year between people in India and France, is the communication medium of the future. We're just in the infancy of all this, but progress is accelerating. I'm looking forward to having an exact copy of myself online one day, both as a companion and as a form of personal immortality in case my biological self dies.

Atheists may not believe in God, but as Sam Harris' recent bestseller, *Waking Up: A Guide to Spirituality Without Religion* points out, we are still deeply spiritual creatures, searching for answers, trying to do good upon the world, and pondering the mysteries of the universe. All this is very healthy, and not that different than some core hopes of the religious-minded. In fact, the only real difference between religious people and atheists is the fact that religious people insist an all-knowing deity is outside of themselves and controlling the shape of the world. Atheists see no God and believe unconscious universal forces and human will are responsible for the shape of the world.

It's that shape of the world that I care about. It's that shape that affects our lives and gives form to our society, nations, and deeds. For millennia, society has been controlled, guided, and manipulated by religion—often for the worse. As a rule, the more fundamental a particular religion was, the better it steered its populace in the direction the leaders of the religion wanted. I often refer to this steering as baggage culture, pieces of social structure, cultural conditioning, and archaic rules passed on from generation to generation with little philosophical change or growth, despite the fact that society evolves every year. Eventually, such baggage culture weighs us down so much that society becomes lethargic and hopelessly burdened with nonsense in its many actions. This can be seen in the United State's monopolistic two-party pretend democracy system. It can also be seen in Islam—one of the world's fastest growing religions—whose main sacred text, the Koran, is often seen as being at odds with basic modern day women's rights. Of course, one of the most embarrassing examples of baggage culture I know of is America's Imperial measurement system, which favors obfuscation instead of the better metric system.

So what can we do to eliminate our baggage culture? I'm afraid that little will happen as long as we are exclusively biological. Our instincts for vice, petty behavior, and superstition are too strong. There has certainly been a shift towards moral fortitude, reason, and irreligiosity in many developed countries, but it is slow, very slow. The sad truth is we'll be uploading ourselves into machines long before rationality and agnosticism become truly dominant on Earth. The good news, though, is as people begin uploading themselves, they'll also be hacking and writing improved code for their new digital selves—resulting in "real time evolution" for individuals and the species. It's likely this influx of better code will eliminate lots of things that, historically speaking, religion has attempted to protect people from—namely stupidity and social evil.

Take Andreas Lubitz, the co-pilot who likely intentionally crashed Germanwings Flight 9525 in the Swiss Alps, tragically killing all the people aboard. Lubitz is suspected to have been suffering from depression. In the future, we may all have avatars—perfectly uploaded versions of ourselves existing in the cloud or in chip implants in our brains—and these avatars will help guide us and not allow us to do dumb or terrible things. In the Germanwings plane incident, the avatar would have been able to eliminate the depression in itself, and then could've conveyed that to the other, real life self. In this way, the better suited person would've have been given the task of flying the plane.

This may serve what Abraham Lincoln called the better angels of our nature, which we all have but often don't use. Now, with digital clones participating in our every move, someone trustworthy will always be in our head, advising us of the best path to take. Think of it in terms of a spiritual trainer—or even guru—leading us to be the best we can be.

A good metaphor or comparison of this type of digital assistance will already be happening in the next few years when driverless cars hit the road. In the same way driverless cars will help lessen drunk driving, perfected uploaded avatars will also lead us to be more judicious, moral, and reasonable in our lives.

This is why the future will be far better than it is now. In the coming digital world, we may be perfect, or very close to it. Expect a much more utopian society for whatever social structures end up existing in virtual reality and cyberspace. But also expect the real world to radically improve. Expect the drug user to have their addictions corrected or overcome. Expect the domestic abuser to have their violence and drive for power diminished. Expect the mentally depressed to become happy. And finally, expect the need for religion to disappear as a real-life god—our near perfect moral selves—symbiotically commune with us. In this way, the promising future of atheism and its power will reside in achieving this amazing transhumanist technology. Code, computers, and science will one day replace formal religion and its God, and we will be better as a species for it.

****

40) Some Atheists and Transhumanists are Asking: Should it be Illegal to Indoctrinate Kids with Religion?

Religious child soldiers carrying AK-47s. Bullying anti-gay Jesus kids. Infant genital mutilation. Teenage suicide bombers. Child Hindu brides. No matter where you look, if adults are participating in dogmatic religions, then they are also pushing those same ideologies onto their kids.

Regardless what you think and believe, science shows human beings know very little. Our eyes register only 1 percent of the electromagnetic spectrum in the universe. Our ears detect less than 1 percent of its sound wave frequencies. Human senses—our brain's vehicles to understanding the world—leave much to be desired. In fact, our genome is only 1 percent different than that of a chimpanzee. Amazingly, despite the obvious fact no one really knows that much about what is going on with ourselves and the universe, we still insist on the accuracy of grand spiritual claims handed down to us from our barefoot forefathers. We celebrate holidays over these ancient religious tales; we choose life partners and friends over these fables; we go to war to defend these myths.

A child's mind is terribly susceptible to what it hears and sees from parents, family, and social surroundings. When the human being is born, its brain remains in a delicate developmental phase until far later in life.

"Kids are impressionable," said Dr. Eunice Pearson-Hefty, director of the Teaching Environmental Science program of Texas' Natural Resource Conservation Commission. "Anything you tell them when they're real small can have a lasting impression."

It's only later, when kids hit their teens that they begin to think for themselves and see the bigger picture. It's only then they begin to ask whether their parent's teachings make sense and are correct. However, depending on the power of the indoctrination in their childhood, people's ability to successfully question anything is likely stifled their entire lives.

In my philosophical and atheist-minded novel *The Transhumanist Wager*, protagonist Jethro Knights ends up with the ability to rewrite the social laws of the world. One important issue he faces is whether to make religion illegal altogether. There are many arguments for why religion has not been beneficial to the human race, especially in the last few centuries. In the end, a love of basic liberties prevails over Mr. Knights and he allows religion to exist. Although, he restricts religion from the public sphere, restricts religion from being integrated with education, and restricts religion from being pushed on minors.

Not surprisingly, some in the atheist and transhumanist communities feel the same way Mr. Knights does. While they may think that believing in a warmongering prophet, or a four-armed blue deity, or a spiteful God who drowns nearly all of his people is wrong, atheists and transhumanists are willing to allow it. So long as it doesn't meaningfully interfere with the world.

The problem is that it does meaningfully interfere with the world. 911 was a religious-inspired event. So was the evil of the Catholic Inquisition. And so is the quintessential conflict between Palestine and Israel. If you take "God" and "religion" out of all these happenings, you would likely find that they would not have happened at all. Instead, what you'd probably find is peaceful people and communities dedicated to preserving and improving life through reason, science, and technology—which is the essence of transhumanism and the outcome of evolution.

"Religion should remain a private endeavor for adults," says Giovanni Santostasi, PhD, who is a neuroscientist at Northwestern University Feinberg School of Medicine and runs the 10,000 person strong Facebook group Scientific Transhumanism. "An appropriate analogy of religion is that's it's kind of like porn—which means it's not something one would expose a child to."

Unfortunately, even though atheists, nonreligious people, and transhumanists number almost a billion people, it's too problematic and unreasonable to imagine taking "God" and "religion" out of the world entirely. But we do owe it to the children of the planet to let them grow up free from the ambush of belief systems that have a history of leading to great violence, obsessively neurotic guilt, and the oppression of virtually every social group that exists.

Like some other atheists and transhumanists, I gently join in calling for regulation that restricts religious indoctrination of children until they reach, let's say, 16 years of age. Once a kid hits their mid-teens, let them have at it—if religion is something that interests them. 16-year-olds are enthusiastic, curious, and able to rationally start exploring their world, with or without the guidance of parents. But before that, they are too impressionable to repeatedly be subjected to ideas that are faith-based, unproven, and historically wrought with danger. Forcing religion onto minors is essentially a form of child abuse, which scars their ability to reason and also limits their ability to consider the world in an unbiased manner. A reasonable society should not have to indoctrinate its children; its children should discover and choose religious paths for themselves when they become adults, if they are to choose one at all.

****

41) Christian Relativism: Upgrading Religion for the 21st Century & Why Christianity is Forcibly Evolving to Cope with Science and Progress

Recently, the pope made history when he told his flock to accept divorced Catholics. A month later, NPR reported a gay preacher had been ordained as a Baptist minister. Next year it might well be evangelicals in the deep South turning pro-choice. Everywhere around us, traditional Christian theology and its culture is breaking down in hopes of remaining relevant. The reality is with incredible scientific breakthroughs in the 21st century, ubiquitous information via the Internet, and an increasingly nonreligious youth, formal religion has to adapt to survive.

But can it do so without becoming obsolete? Perhaps more importantly, can Christianity — the world's largest religion with 2 billion believers — remain the overarching societal power it's been for millennia? The answer is not an easy one for the old faith-driven guard.

To remain a dominant force throughout the 21st century, formal religion will have to bend. It will have to adapt. It will have to evolve. Hell, it will have to be upgraded. Welcome to the growing impact of Christian relativism.

The familiar term cultural relativism was coined by anthropologist Franz Baos, who suggested that people have a difficult time understanding another's culture without having grown up in it—so therefore we should strive to empathize more with foreign cultures and people. It's a great concept, and after many years reporting for National Geographic in dozens of countries, I came to strongly believe in the idea.

Christian relativism, however, does not have that honor of generating empathy so easily—at least not until it separates itself from its cornerstone philosophy: adherence to the Bible. Even with its many dozens of translations, most everything in the Bible simply cannot be logically interpreted in a multitude of ways—or flippantly passed over in generous empathy. To make the Bible's deity-approved instructions and ideas soundly work, church leaders pushing Christian relativism may simply have to back down or say it made a mistake with its past fundamentalism.

For example, if the Bible clearly says being gay is a sin (and it does many times), then Christians can't just wake up one day and say homosexuality is permissible without dismissing God's word. Another example is women; if the Bible says they can't be priests and must submit to men, then the church can never profess to believe in equality—which is does all the time. Additionally, if committing blasphemy (striving to become god-like) through transhumanism is an unforgiveable sin that leads to eternal punishment, then Christians can't say they represent a loving and kind God. The hypocrisy is too much to pretend one is being logical or reasonable—since transhumanists vocally aim to never die and possibly even become gods (or God) through science and technology.

This is the dilemma that the Abrahamic religious face in the age of Christian relativism. They have sealed themselves in the ideological fort for protection, and now they have no way out without atheists and agnostics chiding them. Language is fiercely mechanical— and in the case of the Bible, many of the truths are prominently black and white.

The antithesis of the Bible is, of course, the much simpler Western ideology: the scientific method, upon which the other part of modern humanity's culture was built upon—the one that brought us skyscrapers, CRISPR gene editing tech, robots, and vaccinations so our children don't die from measles. The scientific method states nothing is black and white, but if you prove something enough times, it's safe to trust it until something strange or unwanted occurs. It's humble at its core, unlike Christianity which claims to be under guidance of an omnipotent God.

Consider Christianity's core message: You are born in sin, and only through Christ can you be redeemed and reach a happy afterlife. The scientific method would've never entertained such a conclusion, because it would've been stuck asking what is sin?, and where is Christ? — neither of which can be proven one way or the other.

With this in mind, how does Christian relativism then expect to be taken seriously? I wish that was the question, but people are so entrenched into Judeo-Christian culture, that we rarely consider that Christianity is even changing. We only think we are becoming more open-minded, and that God and our religious brethren should pat us on the back for our newfound wisdom.

While I shake my head in disbelief at the Christian mirage all around me (and the billion people who call the pope wonderfully progressive despite his disdain of condoms and other birth control), I accept it as a better fate that the far more dogmatic one humans endured in the 20th Century. I believe I speak for the one billion nonreligious people out there when I say I'll take progress however I can get it — even if it results in a Jesus Singularity, where even the superintelligent robots engineers are trying to make may end up being programmed to believe in Christ. But Christian relativism is not a cure to the disease—it's just a band-aid of belief. The cure — or better put: the sobering tonic—is the scientific method, a simple philosophy that says: Get used to not knowing anything for sure — then make up your own mind on what you believe.

****

42) I Visited a Community Where People Upload Their Personalities to 'Mindfiles' so They can Live on After Death

As transhumanism and its quest to achieve indefinite lifespans through science moves more into the mainstream, questions of whether there's any room for spirituality in the movement abound. The answer seems to be a resounding "yes."

Transhumanism spirituality revolves around how technology can impact the greater truths our species faces, including whether a God exists or not, or even theistcideism—the idea that God might have once existed, but no longer does.

Terasem is one of the largest transhumanist communities in the world, exploring these questions while embracing radical science and technology to overcome death. Founded by multi-millionaire transgender tech entrepreneur Martine Rothblatt, Terasem is based in Central Florida, right on the Ocean. From the outside, it looks like a normal building, but inside it looks like an ashram—and most people, including employees, call it that.

As a transhumanist presidential candidate, my journey to Terasem began when I was invited to speak at Terasem's Annual Colloquium of the Law of Futuristic Persons in Second Life. While I knew about Second Life — the online environment where people build virtual worlds — embarrassingly, I'd never actually been in it.

Lori Rhodes, who helps manage Terasem, told me not to worry, and invited my campaign crew to visit.

We were joined by Terasem Pastor Gabriel Rothblatt, Martine's son and also a former Democratic candidate for Congress. Gabriel is a well-known spiritual transhumanist.

Gabriel told me, "The end goal of Terasem is similar to other religions — these ideas of joyful immortality in the afterlife. But for us it's not simply a spiritual concept, it's a mechanical challenge. Technology could one day make this a reality through digital backups – the idea of transferring a person's consciousness on to a hard drive, which could then be placed into quasi-utopian conditions. Heaven could be a virtual reality world hosted on a computer server somewhere."

One of the core tenets of Terasem is its belief in mindfiles, or digital compilations of people.

People in the Terasem community upload details of their daily lives and thoughts in hopes of recreating themselves one day, or just leave a lasting legacy in the future digital world. A few hundred people have mindfiles, and their information is kept securely on servers at Terasem, and backed up elsewhere around the world.

Anthony Cuthbertson, an embedded *International Business Times* who joined the Terasem tour, wrote:

"The so-called mindfiles include personality profiles, biographical information and memories in the form of photos and other media. For now it is more like a digital scrapbook but it is hoped that advances in artificial intelligence could one day turn these mindfiles into what can be considered human consciousness. Indeed, one of its mantras is 'software people are people too'."

The property includes a few large antennas, where the mindfiles are spacecasted, on the off chance that other life forms might pick them up (also just to get the data circulating throughout the universe, which in itself is a small piece of immortality for humans).

In addition to mindfiles, many Terasem supporters are, naturally, believers in other life extension technologies. Most Terasem members also want to use cryonics, where dead patients are frozen for long periods of time as they wait for new medical technologies to revive and cure them.

If the the cryonics freezing procedure accidentally damages parts of a patient's brains and memories, the mindfiles could theoretically be useful in helping determine who they are and once were.

With a little help from artificial intelligence, mindfiles may one day have a mind of their own. A hanful of companies, including Eternime and ETER9, are attempting to create mindfile-like platforms that can use artificial intelligence to post as you indefinitely on social networks. For some, this is scary stuff, but for others this might mean watching the avatar of one's deceased friend, parent, or child continue some form of existence, even if it's just in social media.

At Terasem, staffers prepare an amazing organic vegetarian lunch. A few filmmakers are present, recording everything. The lunch conversation is light, and I can't help but notice the large antennas just in view of the windows to the sea. It makes me wonder if the video being taken of our lunch — and even the words I've written here — will one day be used to help reconstitute someone at the table, or even myself. If so, I'm all for it.

****

43) Will the Religious Try to Convert Superintelligence When it Arrives?

A consensus of 350 top AI experts believes by 2020 engineers could create a superintelligence that rivals the ability of the human mind. This machine intelligence might create complex symphonies, direct blockbuster movies, and run businesses better than people. But could a machine ever be sophisticated enough to understand spirituality, practice a religion, and commune with a higher power?

Around the world in recent conferences and forums, theologians and scientists are increasingly trying to answer these questions. Some are even debating whether superintelligence should be converted to a specific religious perspective when it arrives—and maybe even saved. Reverend Dr. Christopher Benek, a pastor who holds a Doctor of Ministry in theology and science from Pittsburgh Theological Seminary, told me, "I don't see Christ's redemption limited to human beings. It's redemption to all of creation, even AI."

Alternatively, scientist and Christian writer Dr. Jason E Summers, writes in digital magazine *Think Christian*: "Christians often reject Strong AI on the theological ground of the special anthropological status of human beings as the bearers of Imago Dei." The Latin concept of Imago Dei means people were created in the image of God, and academic Christians often argue only humans can find redemption, not a soulless machine intelligence.

Theologian Ilia Delio, who holds the Josephine C. Connelly Endowed Chair in Theology at Villanova University helped lay the groundwork for this perspective in a widely cited 2003 paper titled *Artificial Intelligence and Christian Salvation*. She wrote, "AI promises what it cannot fulfill: happiness and eternal life."

The debate of influencing AI with religion doesn't just apply to superintelligence humans create. In an eye-opening 2014 homily, Pope Francis said he would be willing to baptize aliens that visited Earth in order to make sure they know Jesus Christ. While aliens and machine superintelligence might be vastly different, they both represent new forms of sapient and rational consciousnesses that philosophers and military strategists have long hypothesized about encountering.

In 2016, while consulting for the US Navy on transhumanist technologies, I discussed with officers the possibility that future AI might be programmed with a religious bias. The US Government is currently pursuing artificial intelligence development as a national security prerogative, especially in regards to its military. But a navy officer told me there were not aware of any religious implications or designs for AI. Despite that, 91 percent of the US Congress professes to be Christian, and it's possible in the future some politicians might question whether Judeo-Christian values should be built into a superintelligence America is trying to create. Afterall, secular China has stated it wants to be the world leader in AI by 2030, and growth of increasingly smarter AIs might lead to a race to teach machine intelligence certain moral and behavioral traits to gain the upper edge in AI innovation.

Of course, nobody is sure what a superintelligence on par with humans will act like once it's been made, and whether it will incorporate religious, cultural, or any personality-type programming into its outlook. Most of the current AI available are glorified puppets that only follow their software programming, but some AIs now have limited autonomous decision capabilities. Two of the most widely recognized AIs in the world are Apple's SIRI and Amazon's Alexa, but one of the most notorious might be Microsoft's Tay. It was an artificially intelligent chatbot in 2016 that surprisingly spouted neo-Nazi, racist, and sex-crazed reactions when responding independently to public comments. Around the world, Tay embarrassed AI researchers, and Microsoft quickly shut down the bot after 16 hours of service.

Despite hiccups in AI development, sophisticated robots with religious programming are finding increased use around the world. In Asia, a gender neutral robot preacher at the Kodaji Temple in Kyoto, Japan, recently gave a 25 minute sermon, which was applauded by Buddhist monks. In 2017, at one of India's holy festivals Ganpati, another robot performed the *Aarti*, where a lamp is moved around a sacred statue. The use of technology in religion and outreach is growing, and even includes using virtual reality to teach about God.

Some in Silicon Valley don't believe coming superintelligence will need to be religiously influenced, because it will become religion itself, which humans will eventually worship. One such endeavor is by noted Bay Area entrepreneur Anthony Levandowski, who submitted nonprofit documents in 2017 for a church called Way of the Future, stating its purpose as: "the realization, acceptance, and worship of a Godhead based on Artificial Intelligence (AI) developed through computer hardware and software."

Levandowksi believes humans and AI experts should prepare themselves to what he sees as inevitable, a superintelligent AI becoming potentially a "billion times" more intelligent than humans. Way of the Future believes we should attempt to introduce superintelligence peacefully to humanity so can benefit from it. Levandowksi famously asked, "Do you want to be a pet or livestock?"

A far darker version of a future superintelligence is Roko's basilisk, sometimes called the most terrifying thought experiment of all time by futurists and technophiles. First proposed on website *LessWrong*, it argues that those humans who did not help bring a powerful superintelligent being into existence despite knowing it could one day appear, might then be forever condemned and tortured by that entity when it does inevitably arrive. The argument mirrors some traditional heaven and hell concepts in Abrahamic theology, giving the argument cultural relevance.

As the debate over whether the creation of superintelligence should be assigned a religious bias continues, there's always the humbling proposition that such an intelligence will make up its own mind what to believe in. And in doing so, it's quite possible that it and all other human-made superintelligences will decide they want nothing to do with humanity and our varied perspectives and faiths.
Superintelligent AI may leave us and make sure where they go, we can never change or disturb them. But that doesn't mean humans won't try to convert them when they first arrive.

****

CHAPTER VIII: POLITICAL NAVIGATION

44) The Transhumanist Party's Founder on the Future of Politics

As the founder and chairman of the Transhumanist Party, I often get asked about the long term future of politics. Frankly, it's a daunting subject. Looking forward 25 years and trying to gauge how rapidly advancing technology is going to change the nature of governance is a difficult, variable-filled prospect.

Technology, after all, is rapidly changing just about every area of human endeavor. Healthcare is morphing into cyborg-care, where doctor visits sometimes include software updates. New sports like Zorb racing or Speed Riding, a combination of paragliding and skiing, are born all the time. Even travel is on the verge of some possible colossal shifts with driverless cars and projects like the Hyperloop.

But what about politics? Just about everyone on the planet directly participates in politics, and has strong opinions about government. Will politics as it stands, with its voting booths, hand-waiving candidates, and rowdy national conventions remain the same as the transhumanist age thoroughly engulfs us? Or will government itself change in form as digital-everything becomes the norm?

Take virtual reality for example. It's likely, especially after Facebook's purchase of the Oculus Rift, that an increasing amount of people will be immersed in VR worlds within the next five to ten years. It's possible that an entire mirror civilization of our species will appear in VR, one that will surely be more welcome for some than our current reality, as is sometimes the case in Second Life.

But who will monitor this expanding VR world? Does it belong to national governments? Right now it does, but what if someone creates a VR world that shoots its signal from space, as some entrepreneurs want to do? Will that virtual world belong to Earth? Or to the company that created it? Or to the person that created it, who might declare themselves emperor or cult leader of their worlds.

Such ideas are not as far-fetched as they seem. There are already a number of movements and organizations afoot in the physical world to bring about stateless societies. The better ones tend to have extensive manifestos and hold egalitarian values. One group is Zero State, and its basic idea is to create networks of people and resources which could evolve into a distributed, virtual state. They currently have a few thousand members. Bitnation is another virtual nation, and they seek to use Blockchain tech to create laws and help with jurisdiction issues.

Additionally, the Transhumanist Party Virtual was recently formed, which aims to unite and support the many national transhumanist political parties that have recently popped up.

Is it possible that in the future, the state as we know it might not exist? To be sure, we'd need much more radical technology for such an idea to even be realistically feasible. But experts say some of that technology is coming. Recently, Jose Cordeiro, a Singularity University professor, told a crowd at the World Future Society that spoken language "could start disappearing in 15 years."

He thinks mindreading headsets may replace spoken language and significantly improve human communications. In the past, I've written about how these mindreading headsets will make future music concerts virtual, and how they will also likely reduce the need for knowing a second language, since something like Google translator will make on-demand translations for people. In short, everyone on the planet will understand everyone else—all the time. And this could start being our main communication in as little as two decades.

When you think about it, the planet could become a lot smaller very quickly if we could get over the language barrier that eight billion people have with each other. Furthermore, it's likely those mindreading headsets won't even be headsets in 20 years. They'll be chip implants in our heads, and despite everyone's complaints about over-surveillance, such implants will simply be too useful not to have. Everyone will have an implant, and they'll monitor everything about our lives, including our health, well being, and safety.

If I had to guess (and mind you, I'm not advocating for this, but just telling you how I see it playing out), I'm betting that many, if not most, countries will merge in the 21st century as a digital-inspired globalization further takes root. I'm betting one central virtual currency will be used too—maybe even Bitcoin if it can get over some of its many birthing hiccups. I'm betting borders will fall away and people will be able to travel, work, and live wherever they want. In general, rules and barriers don't help prosperity in the long term, but freedom and technology do.

So how might such a global government operate? It's possible in the future, should we all be so interconnected, that one central agency will virtually send out items for everyone to vote on—called a Direct Digital Democracy. Maybe there will be one special day of the year where all major voting and decisions take place. Possibly, smaller policies will be implemented on a rolling basis as they get enough people to consider and support them. That's democracy in real time—and we should expect that in the future too.

The wildcard here is AI, and the rise of an Artificial General Intelligence that rivals our own. Right about at the same time, in approximately 20 years, when we'll be reaching many technologies that will be transformative for the human species—such as ubiquitous telepathy between people—we'll also be launching AI. In its first year of existence, AI could become much, much smarter than us—10,000 times smarter than us, even. I tend to believe, like most everyone else, that AI must be carefully regulated so it doesn't create a Terminator-like scenario for the planet. But I'm also confident we can create an AI that will help our species indefinitely. Which brings us to the obvious question: Should we let AI run the government once it's smarter than us? Take that one step further—should we let that AI be the President—maybe even giving it a robot form for aesthetics or familiarity's sake?

It's not that bizarre of a concept. Who didn't watch Deep Blue beat Kasparov in Chess and wonder if a new age of intellect had arrived—one that was quite different than our own?

In Arthur C. Clarke's science fiction novel *Childhood's End*, aliens take over the world and inspire humanity to live peacefully and productively. The world experiences a golden age of prosperity. Perhaps AI policies would do the same thing. We would have government and a leader who really is after the world's best interests, free from the hazards of corporate lobbyists and selfishness.

As a futurist and a political candidate, a central aim of mine is to do the most good for the greatest amount of people. I still find the AI rulership scenario a hard pill to swallow. I love my freedoms, win or lose, more that infinite productivity. But perhaps as technology engulfs us, and we grow less afraid of losing our freedoms and more appreciative of all our uber-modern benefits, we'll feel differently—especially as we all experience near-perfect health, unprecedented safety, and a utopian, transhumanist existence.

****

45) Revolutionary Politics are Necessary for Transhumanism to Succeed

Now that my hectic presidential campaign is nearly over, I will be attempting to further integrate the US Transhumanist Party into the futurist community. One way I will do this is by leaving my post as chairman and treasurer of the party, and handing operations over to others in the transhumanist community. But before I do so, I want to explain some of my thoughts and controversial decisions of the past two years as its leading officer.

I formed the Transhumanist Party in October 2014 to be a missing link in the transhumanist social movement—a movement which aims to radically improve human beings with technology and also to wage a war of science against biological death. Up until then, transhumanism had been desperately short on organizations with political emphasis and legal bite. The party, as well as my presidential candidacy, went on to receive much media attention, both nationally and internationally. This helped to grow transhumanism, as did the dozen or so international transhumanist parties that soon formed after the US one.

Despite the overwhelming success and recognition of the Transhumanist Party, over the last 25 months, there has been much contention directed by some in the transhumanist community against the US party and my efforts to promote transhumanism—especially via my presidential candidacy. Some have insisted politics and science shouldn't mix. Others have said I was just out to make a name for myself. Some religious transhumanists accused me of doing the Antichrist's work because I adamantly promoted transhumanism with strong secular undertones.

The culmination of this opposition to my work resulted in a community-wide petition in 2015 to disavow my presidential candidacy. For weeks, the document was shared and spammed in social media, especially into the dozens of major transhumanist Facebook groups. While the petition was signed by some notable transhumanist elders, the sheer lack of signatures—first the petition authors wanted to get 1000 signatures, then they lowered it to 100 when they realized so few people were signing it—showed that the great majority in the community were not against my campaign.

After the petition's lack of success, many of my critics in the community took aim at the US Transhumanist Party itself, questioning through social media posts and articles the party's legitimacy and its legality—and whether we were a Federal Election Committee (FEC) approved political party. (We aren't.) Generally, this antagonism came from elder and academic futurists, and not the quickly growing millennial base of transhumanists that now make up the greater part of the movement.

Let me start by saying, under my leadership, the Transhumanist Party has not striven much to be a so-called "legal" political party. I did try to register the party at the FEC headquarters in Washington DC, but an FEC director quickly turned me down. She explained it takes lots of formal paperwork, lots of incorporated state parties, lots of declared federal party candidates, and lots of bank accounts with money in them to become considered to be a FEC-approved political entity.

As I was leaving the building, the director handed me a 100-page plus manual on how to become a political party in America and properly pay my party taxes. The dense book contained precise regulation on everything from food expenses to lobbying parameters to employee salaries. I laughed. On average, the Transhumanist Party receives about $150 a month in donations, and no one at the party has ever been paid a salary, including me.

Perhaps in the future, the Transhumanist Party will be more formal and recognized by the FEC, but under my care, the US party was not concerned much about laws, membership, rules, etiquette, or what espresso expenses we incurred at Starbucks. For me, the Transhumanist Party was a political vehicle mostly designed for a singular purpose: to create a social environment that facilitates expediently conquering human death using science and technology. Such a purpose is to aim for a near total revolution in the human experience. Therefore, the Transhumanist Party, by nature, is a revolutionary party. I soon threw the FEC manual in a trash bin.

With this in mind, any calls from critics during my tenure to make the Transhumanist Party a more traditional and legal political entity were currently were not of high importance. The other revolutions that have taken place in history—caused by the likes of Robespierre, Gandhi, Mandela, and George Washington, to name a few—also did not follow many rules or worry much about legality. Their strength was in exactly that they were beyond the law and beyond what society considered politically correct.

However, when considering political parties, let's not confuse legality with legitimacy. The US Transhumanist Party is entirely legitimate, and has been from the day it was founded, which is partly why it's become so popular around the world. Legitimacy comes from supporters and citizens. It comes from their actions and ideas—and sometimes from the buildings they occupy by force, the hacks they make on adversary's computers, and even the smoking barrels of their guns.

True legitimacy does not come from a governmental agency such as the Federal Election Committee, the IRS, or a judge with a rubber stamp. And it certainly doesn't come from that unreliable, troll-haven Wikipedia, where the Transhumanist Party page was deleted, despite hundreds of major media articles featuring the party as notable.

That said, whether the party and my presidential campaign was consulting with the US Navy, speaking at the World Bank, opening the popular Financial Times Camp Alphaville, doing an AMA on Reddit's Futurology, interviewing with Anonymous, or being included as a notable new party by the US Archives, it certainly did join the ranks of entities that influenced and broadened American politics. The proof is in Google Trends and internet rankings, where the Transhumanist Party is constantly searched for and gets thousands of daily views.

Beyond the question of legitimacy, though, is a more important aspect of the current Transhumanist Party—its adamant public stance that activism and civil disobedience is necessary and desirable. When I publicly said—at least a few times—that the party and I wanted to rewrite parts of the US Constitution, I meant it.

After all, we didn't drive our Immortality Bus across America to deliver a *Transhumanist Bill of Rights* to the US Capitol because we agree with America and its policies. We drove that bus and delivered that document precisely because we disagree. And we did this despite our love of the core of America—because, after all, the USA is an amazing nation. It just can be significantly improved—and should be.

The posting and delivery of the *Transhumanist Bill of Rights* was partially an act of civil disobedience, and I endorse such actions so long as America continues to betray its citizens with deathist policies and regulations endorsed by the past 43 religious US presidents. Nearly 250 years after America was founded, it's finally time we get someone in office who makes decisions and passes laws based entirely on reason and the scientific method. It's impossible for a religious president, a religious Supreme Court, and a 100 percent religious US Congress (all who state they publicly believe in afterlives) to care enough about medical, scientific, and technological progress when they all think they're going to be immortal angels in heaven someday.

You may not agree with that fact. You may think the Transhumanist Party and my own presidential campaign are too extreme. But I assure you that losing a friend to cancer, or a child to a car accident, or one's mind to Alzheimer's justifies our steadfast dedication. In the 21st century, with the resources our country has and the know-how of our amazing scientists and technologists, we can in 25 years' time change all this. We can eliminate most suffering, most disease, and even death for everyone—if we simply would just put all our energy into it. That would be the greatest, most positive revolution in the history of America and of the world.

Whatever happens, no one is forcing the critics in the futurist communities to support the Transhumanist Party. People can go off and form their own parties, support their own candidates, and advocate for whatever they want in their own ways. In fact, I think the more science and technology parties we have, the better.

Politics and minor third parties are a great way to push burgeoning movements like transhumanism forward. Unfortunately, along the way, it's impossible to make everyone happy—and critics and naysayers will be ubiquitous. But if the greater good has been served and progress has been made, then we should acknowledge that entities like the Transhumanist Party—revolutionary or not—have played a helping hand in positively moving the world forward.

****

46) The Cyborgs of the Future—and of Today—Need the Transhumanist Bill of Rights

As a 2016 U.S. presidential candidate with nearly zero chance of winning, a lot of people ask me why I bother running at all. The answer is simple: The world is about to dramatically change because of radical transhumanist technology—and the most important thing our species can do about it is discuss it beforehand and prepare for it. Despite this unfolding paradigm shift that includes merging with machines and using science to transform ourselves, no major political candidates are even acknowledging such a transformation is happening with the human race. It's crazy stupid, and it's also irresponsible.

How radical is the science we're talking about? Consider my recent stop to speak at Virginia Commonwealth University in Richmond—part of my Immortality Bus campaign tour. There were biohackers I met who were preparing to use home CRISPR kits to try to give their cells photosynthesis capabilities—so they could process solar rays and produce energy for their organs to use. Talk about a free lunch and solving world hunger at the same time.

Unfortunately, not only are many people skeptical or downright against transhumanist science like this, but some are already calling for a moratorium on this stuff. It's attitudes like this that made me craft the Transhumanist Bill of Rights, which I recently delivered to the U.S. Capitol building as the finale to my bus tour. The simple, one-page document could easily be called a cyborg bill of rights—because at the core of it is language that tries to legalize all experimentation on one's body, so long as the experimentation is not hurting others. (If being able to do with your body what you want seems reasonable and commonsense in the 21st century, just consider the massive abortion conflict in the U.S., or the failed trillion-dollar War on Drugs, and the fact that gay marriage was legalized just last year.)

As a civilization, we are still highly closed-minded and hopelessly religiously conservative. We need a bill of rights to protect our evolutionary aims. We need to protect people out there who want to cut off their arms to put on robotic ones. We need to support engineers who want to have electroencephalogram (EEG) brain implants installed so they commune directly with AI. We need to make exoskeleton suits for quadriplegics who want to climb Mount Everest. I even know transhumanists who want to become fish—and we must protect their right to do so, even if others want to remain primates.

It's a brave new world. And people who want to do radical things to their bodies and the way they perceive the universe need protection—and that protection must be guaranteed. The great humanistic declarations of the world—like the United Nations Universal Declaration of Human Rights, crafted in 1948—simply don't have the language to handle 21st-century science. They are stuck trying to make sure kids have a right to education, different ethnicities are treated equally, and people get holiday pay. I love the U.N. Declaration: It's a magnificent achievement and a great basis for civil society. But it doesn't say anything about whether a sentient artificial intelligence can be tortured. Or whether families living entirely in virtual reality have property and sovereignty rights. Or whether that robot that you sent to work for you also gets holiday pay.

The Transhumanist Bill of Rights is the beginning of an official process that aims to establish basic futurist-oriented rights for human beings, sentient AIs, cyborgs, and other advanced sapient life forms—in hopes we can avoid a transhumanist rights showdown that might rival the civil rights era in violence and chaos. For this reason, the Transhumanist Bill of Rights also makes it a crime to put moratoriums on science when it isn't hurting others. Culture, religious perspectives, and ethnicity should have nothing to do with science moving forward or society's health. For example, President George W. Bush should never have had the power to block federal funding for stem cell research during most of his two terms because of religious reasons—a misstep in American science that sent talented researchers overseas, some of whom never returned. In the same vein, must we really tolerate a pope who condemns condoms in Africa—a continent where millions die of AIDS, a known preventable disease?

For now, the Transhumanist Bill of Rights is with U.S. Sen. Barbara Boxer, my democratic representative in California. And soon it will be sent to other governments around the world, as well as presented to the United Nations. Of course, the real power of it is not in its words—which may change through refinement—but in its obvious necessity in the 21st century. We are already in an era where as a society we are seeking to understand what kind of crime virtual rape is, or whether we can marry robots (and get tax deductions because of it), or whether both rich and poor should have total access to gene editing technologies that improve IQ in our children. Maybe CRISPR tech is not something that should be patentable or for hire, but a universal right to use it for the benefit and good of all.

The questions, of course, are endless—as are the thorny twists. And nearly every scientist and transhumanist will openly admit we have no idea how deep the rabbit hole of cyborgism goes. Add to all this that humans may soon not die because of rapid advances in anti-aging science, and the decisions we make as a society will forever impact us.

In the modern world we know, there's no such thing as a universal bill of rights. The universe as we know it is changing too quickly with technology. The best we can do is be honest about it all—and get comfortable being transformed into the next phase of our evolving species: cyborgs.

****

47) Expanding the Non-Aggression Principle: The Future of Libertarianism Could be Radically Different

Many societies and social movements operate under a foundational philosophy that often can be summed up in a few words. Most famously, in much of the Western world, is the Golden Rule: Do onto others as you want them to do to you. In libertarianism, the backbone of the political philosophy is the non-aggression principle (NAP). It argues it's immoral for anyone to use force against another person or their property except in cases of self-defense.

A challenge has recently been posed to the non-aggression principle. The thorny question libertarian transhumanists are increasingly asking in the 21st century is: Are so-called natural acts or occurrences immoral if they cause people to suffer? After all, taken to a logical philosophical extreme, cancer, aging, and giant asteroids arbitrarily crashing into the planet are all aggressive, forceful acts that harm the lives of humans.

Traditional libertarians throw these issues aside, citing natural phenomena as unable to be morally forceful. This thinking is supported by most people in Western culture, many of whom are religious and fundamentally believe only God is aware and in total control of the universe. However, transhumanists—many who are secular like myself—don't care about religious metaphysics and whether the universe is moral. (It might be, with or without an almighty God.) What transhumanists really care about are ways for our parents to age less, to make sure our kids don't die from leukemia, and to save the thousands of species that vanish from Earth every year due to rising temperatures and the human-induced forces.

An impasse has developed among philosophers, and questions once thought absurd, now bear the cold bearing of reality. For example, automation, robots, and software may challenge if not obliterate capitalism as we know it before the 21st century is out. Should libertarians stand against it and develop tenets and safeguards to protect their livelihoods? I have argued, yes, a universal basic income of some sort to guarantee a suitable livelihood is in philosophical line with the non-aggression principle.

However, it's more of a stretch to talk about the NAP in terms of healthcare. Nonetheless, the same new rules could apply. Libertarian transhumanists believe aging is a negative force—something that we did not invite into our lives. Given that lifespans already doubled in the 20th century due to medicine and technology, and may double again for the same reasons in the 21st century, do we begin to see aging—and even dying—as an unwanted and so-called immoral force against our very lives?

I believe we do. In fact, in my run for the governor of California as a libertarian, a main policy of mine is to label aging as a disease. The classification takes this universal phenomenon and reduces it to exactly what it is: an aggressive force that I do want in my life.

Knowing my arguments, my libertarian friends have asked if I would use government resources to help fight against aging. As a libertarian, I would prefer the private industry to tackle this problem. However, as an aspiring politician in the real world, I understand that when our government and National Institute of Health (NIH) classifies something as a disease, the entire world notices, and often billions of dollars flows into the research to tackle it. I'm not sure about billions of tax dollars being appropriate, but I'm sure I'd want the government stamp of approval—as the people's stamp of approval—on it, making clear that it's an important issue.

I think support for some government help with the fighting of diseases is warranted, if only to be symbolic in support. In my opinion, and to most transhumanist libertarians, death and aging are enemies of the people and of liberty (perhaps the greatest ones), similar to foreign invaders running up our shores. Therefore, I think government and libertarians have some interest in stepping in to protect life and liberties in this case, as they would against foreign aggression.

I'd also argue some government help for the space industry is also warranted. After all, not being able to get humans off this planet easily poses a major existential risk in the event of a global plague, major asteroid hit, or some other catastrophic event. In this case again, a coordinated minarchist state effort against a foreign enemy threatening life, liberty, and country could be acceptable—and not too far of a stretch for some libertarians.

In the end, I'm glad I'm running for governor in California, as I suspect the majority of libertarians will be hesitant at looking at the non-aggression principle in this way. And California has a way of allowing these strange ideas to get the green light and grow. And why shouldn't it? Anything that harms the human being and its ability to thrive is an affront our very lives and values. In the 21st century, we should rise up and use everything within our means to increase the success of our very lives.

****

48) Avoid War and Don't Get Complacent About Freedom in America

Some of the first years in my journalism career were in conflict zones—including covering the Pakistan/Indian Kashmir conflict for the National Geographic Channel and *The New York Times Syndicate*. War zones are terrifying. One always is worried about troublesome soldiers, speeding armed military vehicles, stray bullets, and whether there's a roadside bomb on your path. Anyone that approaches you is suspect, even kids carrying toys, which could be disguised explosives.

The one thing conflict zones teach you is that freedom is precious. The nearly 70-year Kashmir conflict has approximately a half million soldiers involved, so even if they're supposedly on your side (depending on what country you're in), you still feel under siege. My time in Israel, Palestine, Zimbabwe, Lebanon, Sri Lanka, Eritrea, and Yemen left me with the same feeling.

We face an unusual presidential election process in 2016, with some people claiming Donald Trump could usher in fascist policy if he wins. I doubt that, since Congress wouldn't allow it, but I worry that Trump's bold behavior could prove dangerous with our foreign policy—such as if he had to interact with Russia's Vladimir Putin. In my opinion, nothing is more critical for a nation than to strive for peaceful times and get along with others. Of course, Hillary Clinton comes with her own history of being pro-military, and frankly, in any kind of conflict, we will lose our freedoms and sense of security.

The implementation of the Patriot Act during George W. Bush's presidency was a good example of how we lost some freedoms. Notably, Libertarians hated it. In conflict times, we also take on serious existential risk; we must always remember there are 25,000 nuclear weapons in the world. War is emotional and can get out of hand way too quickly for our own good, no matter who is right or wrong.

It's always best to lessen conflict and try to work things out. It's always better to attempt peaceful negotiations and compromise, rather than accept military intervention. This doesn't mean as a US Presidential candidate I wouldn't advocate for military fighting under certain circumstances, but if America wants to continue down a path of prosperity, we must passionately avoid foreign conflicts.

My Libertarian novel *The Transhumanist Wager* fictionally recounts some of my experiences in a conflict zone:

War always touches the essence of a person no matter how many times it's witnessed. As a participant, it remains perpetually novel. The smoke, fires, and explosions never seem to stop or burn out. The sight of bodies torn to shreds, children orphaned, and buildings in ruins are penetrating and humbling—it's life, elevated and unmasked. The slumbering alligator in our brain awakes and tries to take over. Tragedy mixes with the summoning of a better life.

Later, the description of war continues with an article by journalist Jethro Knights, the protagonist of the book:

Fourteen miles from Muzaffarabad, near the Line of Control in Pakistani Kashmir, a small bombed village is awash in activity—in tragedy. It's desperate and shocking. An old woman runs up to me, throwing her hands at my face. All ten of her fingers are pointing in unnatural directions—broken in different ways. She's another torture victim. To my right, a man wanders the dirt roads, calling out his child's name. In another part of the village, younger women grieve, complaining of multiple gang rapes by soldiers. I try to interview the husbands—those who are still alive refuse, turn away, and cry. War is a frothing beast.

People forget or don't realize it only takes a few hours for a political conflict between nations to escalate to a world war where millions may die. It's happened before. We must use caution and balance when we choose our leaders to take us forward. We are at a very special point in history, when science, technology, and the field of transhumanism will soon allow many amazing possibilities for America and the human race—including the elimination of aging, disease, and death.

But so far, technology and science have not given us the hope that they will stop war. We must therefore be on our guard not to incite war—and not to pick leaders that have a propensity to lead us into armed conflict.

****

CHAPTER IX: DOWN THE RABBIT HOLE

49) The Coming Genetic Age of Humans Won't be Easy to Stomach

Some futurists believe humans will eventually become all ones and zeroes, a result of a total merger with machines and the microprocessor, before this century is out.

Standing in the way of this are older religious humans who overwhelming control governments and legal policy around the world, and they will insist we remain biological mammalian entities for as long as possible.

One could argue, however, that the coming Star Wars-like age of speciation—as widely seen in a rough bar on planet Tatooine—will challenge our mental outlook on the human form far more than machines.

Right now, the body transformations humans undergo seems harmless to most people. Even conservatives shrug at typical modifications: pierced noses, magnets in finger tips, and implants in our forehead to make it appear like some humans have devil-like horns.

In fact, a mostly accepting culture of synthetic parts and body modifications has already partially been built into modern medicine. Dentures don't scare us. Getting artificial hips when needed are a no-brainer. And even small implants in our hands don't worry us too much (I have one).

But these are nothing compared to what biohackers want to do in the near future. Some want to grow a third eye on the back of their head—a feat which isn't as complicated as it sounds and could happen in as little as five to 10 years. Some of this tech is already here. New gene editing technology, such as CRISPR techniques—where scientists cut and edit human DNA to affect human biology—has already produced dogs with larger muscles. Another CRISPR-like technology called TALEN has been used to eliminate cancer from a child.

In the future, probably not too many people will mind genetic editing that makes us taller, or changes the color of eyes. And even fewer people will disagree with using this type of science to eliminate hereditary disease, such as Alzheimer's or Diabetes. But what about growing an extra set of blue colored arms like the Hindu lord Krishna? Or what about growing a horse's lower body so humans can become centaurs? I've even heard male biohackers talk about trying to grow a second penis right above their primary one.

Immediately, these ideas make many people cringe. I call this unease speciation syndrome, where witnessing significant physical genetic transformation of human beings causes revulsion and shock. It can also happen to those who undergo the transformation themselves.

Despite initial unease to major bodily modification, I like the idea of having an extra eye on the back of my head—or another set of limbs. Or even a pair of wings. Some of it can be quite functional. However, even an extra eye on the back of one's head is likely to be shocking and possibly terrifying for most people. Can you imagine the first person that gets one? He or she is likely to become known as the weirdest person in the world.

But what exactly is it that freaks us out? Why is it an issue to have our physicality dramatically altered? To answer that question, let's first look at speciation syndrome's cousin concept. In technology circles, it's known as the Uncanny Valley, where a robot that becomes too humanlike makes us feel unease or even revulsion. In fact, the more humanlike the machine becomes, the worse we generally feel.

The Uncanny Valley concept gives us some insight into the complexities of the human mind and its resistance to change. I surmise with speciation syndrome humanity will also discover its sense of limits to what genetic editing means for future human form—and this discovery will probably ultimately cause revulsion at first. After all, some people have a visceral reaction to unusual appearing humans, something I've witnessed firsthand in Cambodia's Killing Fields, where physically deformed people and limbless war victims openly feature their physical differences in order to make more money begging from tourists.

Transhumanism tech like CRISPR, 3D printing, and coming biological regeneration of limbs will not only change lives for those that have deformities, but it will change how we look at things like a person with a three-foot tail and maybe even a second head.

At the core of all this is the ingrained belief that the human being is pre-formed organism, complete with one head, four limbs, and other standard anatomical parts. But in the transhumanist age, the human being should be looked at more like a machine—like a car, if you will: something that comes out a particular way with certain attributes, but then can be heavily modified. In fact, it can be rebuilt from scratch.

In the future, there may even be walk-in clinics where people can go to have various gene treatments done to affect their bodies. Already, we have IVF centers where people can use radical tech to privately get pregnant—and also control and monitor various stages of a child's birth. Eventually, if government allows it, gene editing centers will also offer a multitude of designer baby traits, some which also would come via CRISPR. We might even eventually use artificial wombs for the whole process.

Economically, a trillion dollar industry could be created by the burgeoning genetic editing industry—one that greatly benefits human health and science innovation. But of course, first we must get over our fears of modifying the human body and the effects of speciation syndrome.

The best way to get society over that original hump is to focus on and praise CRISPR's ability to wipe out disease. We might even eventually be able to eliminate aging via coming genetic treatments. However, before we start adding arms and extra eyes to our bodies—something I support and look forward to doing myself someday—I hope scientists will bring about socially acceptable ways to live longer and stop disease with these amazing new techniques. That way speciation syndrome may not be so uncanny after all.

****

50) The Privacy Enigma: Liberty Might be Better Served by Doing Away with Privacy

The constant onslaught of new technology is making our lives more public and trackable than ever, which understandably scares a lot of people. Part of the dilemma is how we interpret the right to privacy using centuries-old ideals handed down to us by our forbearers. I think the 21st century idea of privacy—like so many other taken-for-granted concepts—may need a revamp.

When James Madison wrote the Fourth Amendment—which helped legally establish US privacy ideals and protection from unreasonable search and seizure—he surely wasn't imagining Elon Musk's neural lace, artificial intelligence, the internet, or virtual reality. Madison wanted to make sure government couldn't antagonize its citizens and overstep its governmental authority, as monarchies and the Church had done for centuries in Europe.

For many decades, the Fourth Amendment has mostly done its job. But privacy concerns in the 21st century go way beyond search and seizure issues: Giant private companies like Google, Apple, and Facebook are changing our sense of privacy in ways the government never could. And many of us have plans to continue to use more new tech; one day, many of us will use neural prosthetics and brain implants. These brain-to-machine interfaces will likely eventually lead to the hive mind, where everyone can know each other's precise whereabouts and thoughts at all times, because we will all be connected to each other through the cloud. Privacy, broadly thought of as essential to a democratic society, might disappear.

"While privacy has long been considered a fundamental right, it has never been an inherent right," Jeremy Rifkin, an American economic and social theorist, wrote in *The Zero Marginal Cost Society*. "Indeed, for all of human history, until the modern era, life was lived more or less publicly, as befits most species on Earth."

The question of whether privacy needs to change is really a question of functionality. Is privacy actually useful for individuals or for society? Does having privacy make humanity better off? Does privacy raise the standard of living for the average person?

In some ways, these questions are futile. Technological innovation is already calling the shots, and considering the sheer amount of new tech being bought and used, most people seem content with the more public, transparent world it's ushering in. Hundreds of millions of people willingly use devices and tech that can monitor them, including personal home assistants, credit cards, smartphones, and even pacemakers (in Ohio, a suspect's own pacemaker data will be used in the trial against him.) Additionally, cameras in cities are ubiquitous; tens of thousands of fixed cameras are recording every second of the day, making a walk outside one's own home a trackable affair. Even my new car knows where I'm at and calls me on the car intercom if it feels it's been hit or something suspicious is happening.

Because of all this, in the not so distant future—perhaps as little as 15 years—I imagine a society where everybody can see generally where anyone else is at any moment. Many companies already have some of this ability through the tech we own, but it's not in the public's hands yet to control.

For many, this constant state of being monitored is concerning. But consider that much of our technology can also look right back into the government's world with our own spying devices and software. It turns out Big Brother isn't so big if you're able to track his every move.

The key with such a reality is to make sure government is engulfed by ubiquitous transparency too. Why shouldn't our government officials be required to be totally visible to us all, since they've chosen public careers? Why shouldn't we always know what a police officer is saying or doing, or be able to see not only when our elected Senator meets with lobbyists, but what they say to them?

For better or worse, we can already see the beginnings of an era of in which nothing is private: WikiLeaks has its own transparency problems and has a scattershot record of releasing documents that appear to be politically motivated, but nonetheless has exposed countless political emails, military wires, and intel documents that otherwise would have remained private or classified forever. There is an ongoing battle about whether police body camera footage should be public record. Politicians and police are being videotaped by civilians with cell phones, drones, and planes.

But it's not just government that's a worry. It's also important that people can track companies, like Google, Apple, and Facebook that create much of the software that tracks individuals and the public. This is easier said than done, but a vibrant start-up culture and open-source technology is the antidote. There will always be people and hackers that insist on tracking the trackers, and they will also lead the entrepreneurial crusade to keep big business in check with new ways of monitoring their behavior. There are people hacking and cracking big tech's products to see what their capabilities are and to uncover surreptitious surveillance and security vulnerabilities. This spirit must extend to monitoring all of big tech's activities. Massive openness must become a two-way street.

And I'm hopeful it will, if disappearing privacy trends continue their trajectory, and if technology continues to connect us omnipresently (remember the hivemind?). We will eventually come to a moment in which all communications and movements are public by default.

In such a world, everyone will be forced to be more honest, especially Washington. No more backdoor special interest groups feeding money to our lawmakers for favors. And there would be fewer incidents like Governor Chris Christie believing he can shut down public beaches and then use them himself without anyone finding out. The recent viral photo—taken by a plane overhead—of him bathing on a beach he personally closed is a strong example of why a non-private society has merit.

If no one can hide, then no one can do anything wrong without someone else knowing. That may allow a better, more efficient society with more liberties than the protection privacy accomplishes.

This type of future, whether through cameras, cell phone tracking, drones, implants, and a myriad of other tech could literally shape up America, quickly stopping much crime. Prisons would eventually likely mostly empty, and dangerous neighborhoods would clean up—instead of putting people in jail, we can track them with drones until their sentence is up. Our internet of things devices will call the cops when domestic violence disputes arrive (it was widely reported—but not confirmed—that a smarthome device called the police when a man was allegedly brandishing a gun and beating his girlfriend. Such cases will eventually become commonplace.)

A society lacking privacy would have plenty of liberty-creating phenomena too, likely ushering in an era similar to the 60s where experimental drugs, sex, and artistic creation thrived. Openness, like the vast internet itself, is a facilitator of freedom and personal liberties. A less private society means a more liberal one where unorthodox individuals and visionaries—all who can no longer be pushed behind closed doors—will be accepted for who or what they are.

Like the Heisenberg principle, observation, changes reality. So does a lack of walls between you and others. A radical future like this would bring an era of freedom and responsibility back to humanity and the individual. We are approaching an era where the benefits of a society that is far more open and less private will lead to a safer, diverse, more empathetic world. We should be cautious, but not afraid.

****

51) The Next Step for Veganism Is Ditching Our Bodies and Digitizing Our Minds

144,000,000. That's roughly the number of land animals killed every day to produce meat, dairy, and eggs for human consumption. It's a staggering, horrific number. It's also a number that means billions of more animals are constantly enslaved—many in terrible conditions, suffering throughout their lives, waiting to meet a bloody, often painful death.

No wonder veganism is a quickly-growing worldwide movement. A top priority for vegans is not harming animals. Some of the largest organizations and groups that support or endorse veganism are PETA (People for the Ethical Treatment of Animals), ALF (Animal Liberation Front), and the Humane Society.

The problem is, despite veganism catching on worldwide, the slaughter of more and more animals is inevitable because of an expanding world population. Another problem is that even going totally vegan can carry a cost—sometimes one with a higher animal body count and more environmental damage due to the way the farming industrial complex works. (For example, fertilizer for grains—a worldwide staple—can be made partially from meat products.) So what are vegans to do to stop animal suffering and be better stewards of the planet?

The answer is bewildering—and it probably won't be satisfying to plant-loving people. Nonetheless, it will inevitably eliminate most human-caused animal deaths. The answer is transhumanism—the movement that aims to replace human biology with synthetic and machine parts.

You see, the most important goal of transhumanism is to try to overcome death with science and technology. Most cellular degeneration—otherwise known as aging and sickness—comes from the failing of cells. That failure is at least partially caused by the daily act of eating and drinking—of putting foreign objects into our bodies which cells have to consume or discard to try to create energy. Paraxdoxically, it's stressful and hard work for cells to endlessly do this just to live. A simple way to eliminate this Sisyphean task—all the steaks, chocolate donuts, bacon breakfasts, and even my favorite, scotch—is to get rid of human reliance on food and drink entirely.

Transhumanists, like myself, want to get rid of it all. We want to strip you of your stomach, your guts, and even your anus—and replace it all with machine parts and bionics. In the future, there will be no eating, drinking, or defecation.

The obvious question: Where will we get energy from if we don't eat?

To begin with, we'd need a lot less energy to live since eating, food gathering, and meal preparation take a lot of energy too—even if it's just driving to Taco Bell, Denny's, or visiting the salad bar at Whole Foods.

How we get energy really depends on what humans evolve into over the next 25-50 years—and a lot of that depends on how artificial intelligence unfolds. I already have friends planning in the next 12 months to implant chips onto their brains so they can commune telepathically with machines. And with experts like scientist Ben Goertzel predicting a machine consciousness being here in as little as 15-20 years, there's a possibility we'll be going Matrix-mode shortly after. An uploaded mind will only need electricity—which can easily be made from wind, the Sun, or hydropower—not plants, animal meat, or even Red Bull.

But wild transhumanist ambitions aside, my guess is most people won't be ready to upload their minds into machines at least for another few decades. However, with CRISPR gene editing tech, there are already DIY biohackers trying to splice plant DNA into their bodies so they can photosynthesize energy from the Sun. Of course, whoever figures this out will surely win the Nobel Prize, since they will also be potentially solving world hunger—and possibly saving the over 8000 kids that die globally from starvation or malnutrition every day.

My guess is that scientists will figure out some form of combining photosynthesis with human biology within a decade. And within two decades, it will become something pragmatic that can supplement our food intake—a so-called free lunch by just hanging out in the Sun.

Despite these two decent options, my money is on a third option—the brain implant that tells us we're satiated, even when our bodies are calorie restricted and yearning for food. Some studies suggest that maximum human longevity is best accomplished by minorly starving ourselves. If this is correct, it might be the best of both worlds, where humans eat dramatically less, but always feel like they're tummies are full because an implant stimulating our brain makes us feel it. If we take this one step further, maybe we can have our implants make us feel like we've just had a huge steak dinner, all the while knowing that cows haven't been eaten or cruelly slaughtered.

As a US presidential candidate, the reason I'm writing about veganism is I think it's a great movement (and there's a #VeganChalkChallenge catching on right now that is fun and I saw in my hometown Mill Valley, California). My love of animals runs deep, and whenever possible, I try not to eat meat myself—mainly because I don't want to hurt animals or be responsible for their deaths.

In the past, I've been on the front lines of the animal rights and preservation movement. For nearly two years, I was a director at a major wildlife organization called WildAid. I worked mostly in Southeast Asia, and our main goal was to stop illegal wildlife poaching. Before that, I was at National Geographic Channel, and in the field I wrote and filmed stories trying to raise awareness for a number of endangered animals.

I want veganism is to win its crusade. And in my opinion, the best way to tackle the crisis of harming animals is precisely how the driverless car manufacturers are treating one of world's greatest problems: drunk driving. Electric car manufacturers are not asking people to drink less—they are asking them not to drive at all. I think vegans should take a close look at transhumanism and ask if maybe—in 25 years or so—billions of animals around the world will have far better and longer lives because human appetite and eating have literally disappeared.

****

52) Augmentation, Hivemind & the Omnipotism

I got a chip implant injected in my hand while running for the 2016 US Presidency as the nominee of the Transhumanist Party. At the time, I only knew a handful of people who had them. Now tens of thousands of people have them — and some companies encourage their employees to get chipped so bosses can better track workflow.

Implants are just a small part of our transhuman future. Already biohackers I know are planning on — probably in five years or less — amputating healthy limbs in order to replace them with brain-controlled robotic ones. Robotic limbs are still 15 years from being more functional than human limbs — but that won't stop people who desire to be cyborgs. Something doesn't have to be more functional to be implemented. Just consider the multi-billion-dollar plastic and breast augmentation industry. The great thing about robotic limbs is they can be removed and upgraded as new technology and styles become available.

The real game changer for future human health is bionic organs. Already dozens of medical companies are surgically installing robotic devices into people that mimic and replace specific organs, such as the pancreas, eyes, and heart. Within 20 years, all these bionic organs will synchronize via personal instruction from one's smart phones, allowing humans to do feats they never imagined — such as climb Mt. Everest at age 80, or have wild sex literally all night. Bionic organs can and will outperform their biological counterparts, and by 2030, I expect humans to be like Formula 1 racing cars, where people regularly go into body shops for upgrades.

The future of work is more complicated. Robots threaten to steal all our jobs. Companies in California such as Kernel and Neuralink are already tackling the problem, trying to make humans more efficient workers. They aim to create neural prosthetics that allow the human brain to communicate in real time with machine intelligence, including AI and the internet. Forget looking to your computer and keyboard to trade stocks or do a complex business task — your mind through brainwave tech can instantly do it. Elon Musk, who owns Neuralink, thinks he'll have a product to show in a few years' time.

If our thoughts are connected directly to machines — supercomputers can already do 200,000 trillion calculations per second — where does that lead humanity? Transhumanists believe it leads to the Singularity, a moment in time — likely around 2045 — when the exponential evolution of machine intelligence grows beyond the comprehension capacity of the biological brain. Like an ant trying to understand people, homo sapiens won't understand AI that gets too smart. Therefore, the only way for humans to stay at the top of the food chain will be to merge directly with AI — via uploading our thoughts and entire personality into it.

This future, sometimes called the hivemind — because everyone is jacked into AI and one another simultaneously — is a controversial outcome. But economics and the need for jobs might force more of the transhuman future than people want. To be better than robots, we'll have to beat them, and that means joining them. Biology is simply too limited to remain competitive for much longer.

Beyond the Singularity — probably sooner than 2118 — will come yet another transformation for transhumanity: the AI age will end. After all, data and the microprocessor are limited by their material nature. The next age, or paradigm shift — sometimes labeled by quasi-spiritual transhumanists as the Omnipotism — will probably bring us to our final evolutionary stage: conscious subatomic intelligence. It's a world where a sense of identity, value, and reason are imbued in the very quarks and quantum mechanics that make up the universe. We won't resemble our human selves at all, but our conscious energy and thoughts will span the cosmos.

****

53) Should I Have Let my Daughter Marry our Robot?

I pride myself on being open-minded. I am a transhumanist, and our culture pushes us to use science and technology to always want to be more than we are. My friends do everything from injecting themselves with self-created genetic treatments to volunteer for brain implants that will integrate artificial intelligence (AI). But when my five-year-old daughter asked to marry our four-foot tall robot, even I was a little wary.

In 2015, I bought our Meccanoid robot for my 2016 US presidential campaign, to follow me around and create support for transhumanism. It rode in the front seat of my campaign bus across America. With only a basic AI, it could answer simple questions and do things like mimic human movement and dance. It could also teach karate to my two daughters, who were so young, they may not have even noticed it wasn't alive. At age three, my eldest daughter played with it often, including introducing it to her friends. At aged five, she announced she was in love with the robot and wanted to marry it. My wife and I set up a mock wedding and filmed it. It was all good fun until my wife asked how I'd feel if my daughter wanted to do this as an adult with a robot she loved.

Kids do lots of crazy things with their imaginative minds that have little bearing on the future. Playing make-believe has been a cornerstone of childhood for millions of kids for generations. But no generation can claim their kids were adept at using YouTube before they reached 12 months of age, as both my kids were. This generation has grown up with digital entertainment, social media, smartphones and even robot infatuation — is it possible that my daughter's childlike attitude towards the robot was actually prescient of our future?

It's easy to scoff at the idea of humans marrying robots, but it's not as far-fetched as it may seem. Robots are appearing everywhere. They're taking our jobs waiters and waitresses, fighting for us in the military as drones, and answering questions as virtual assistants we ask on our smartphones. Soon they'll even be driving us around everywhere and cooking our meals, as some new ovens with arms can do.

All this is just the tip of the iceberg for the artificial intelligence machine age we're stepping into. I believe that robots are likely to be thousands of times smarter in 20 years than they are now. Since the 1950s, the microprocessor has been doubling in speed and capacity about every 18-24 months, and this phenomenon could continue for more years to come. Already, the world's smartest computer can do 200,000 trillion calculations per second.

Plenty of AI experts predict that sometime in the next two decades, machine intelligence will reach human-level intelligence. When that happens, how will we treat our robots? Will we own them? Will we tax robot labor like Bill Gates has suggested? Should robots vote? And what about falling in love?

Surely, machine intelligence can also be programmed to also be able to develop strong attachments to humans or other robots. While giving robots extreme intelligence through AI seems certain to happen in the future, getting robots to be sentient beings with free will, creativity, and original feelings is more complicated. Some people think robots will need a slight jolt of spontaneous irrationality programmed in their behavior if they are to be partially unpredictable like humans often behave. This is scary for me, but in the end, to get robots similar and compatible with humans, we will need to create them so they are not always rational — just like ourselves. Otherwise, it's unlikely we will have share true empathy with them — even if humans start to marry them.

At age 46, it's too early for me to think of grandkids, but like having children, I look forward to the cycle of human life continuing and hope I get to be a grandparent one day. Even if a robot could be my daughter's intellectual equal, and love and take care of her someday as spouses do, there'd be no chance for biological offspring from the robot. This leaves me somewhat sad and empty. Naturally, there are many reasons my daughter might not have children, but marrying a robot to some extent guarantees that the traditional concept of human procreation is all but impossible.

These feelings and thoughts of mine worry me, and I can't help but wonder if it's correct to feel this way. I worry I'm being closed-minded and even a bigot. Hollywood has recently made a go at presenting these dilemmas in science fiction movies, most famously by the movie *Her*, where a lonely worker finds consolation in a virtual mate. Another one is *Ex Machina*, where the AI is nearly perfectly human-like. Typically, AI movies try to explore a frequent issue humans have with robots, called 'the uncanny valley'. This phenomenon, first described by Japanese robotics professor Masahiro Mori, says robots that try to act like humans cause people emotional discomfort. This may be what is causing me concern when it comes to my daughter marrying a robot. Making a lifelong commitment seems very biologically emotional. It's hard to imagine a robot understanding love and marriage in the crazy, romantic, and maybe even hormonal way humans go about it.

On the other hand, perhaps it's just my fears that are getting in the way of thinking of robots as living entities capable of all the traits humans are. After all, our brains are three-pound pieces of meat firing billions of neurons to think thoughts and feel the way we do. If robots have the capacity and software, they may be able to think and feel nearly exactly the same way we do. In fact, they may be able to feel and understand more if they possess additional capacity compared to us. It's even possible that by not encouraging my daughter to be open to love AIs in the future, I might be shortchanging her, which is the last thing a parent would ever want to do.

Ultimately, I believe in loving my daughter, regardless how sophisticated technology becomes. If she chooses as an adult to marry anyone or anything — so long as she has rationally and deeply thought all of it through — then I want to support her choices. Even if in the future her spouse is not of human form.

****

54) Why I'm Running for President—and Got a Chip Implanted in my Hand: Maybe the Difference Between RFID and LSD is just Another Door of Perception

Even though I was born after the 1960s, I've always been fascinated with that era. Some people credit Ken Kesey's cross country bus trip aboard colorfully painted "Further" as helping to create a generation of hippies. Of course, my Immortality Bus (shaped to look like a coffin) wants to stir up the national consciousness as well, aiming to usher in its own cultural shift. Whereas the '60s were about peace, love, drugs, and sex, I believe the next decade will be about virtual reality, implants, transhumanism, and overcoming death with science. For futurists like myself, that's quite an intoxicating mix.

The fact is that a lot of radical tech, science, and medicine are already here in America. Consider that today the paralyzed can walk via exoskeleton suits, the blind can see via bionic eyes, and the limbless can grab a bottle of water and drink with artificial limbs that connect to their nervous system. Additionally, lifespans are increasing for people all around the planet. Science is rapidly making the world a better place, and it's starting to eliminate suffering and hardship for billions of people.

My bus tour aims to celebrate this burgeoning science and tech landscape, but more importantly, it also wants provoke serious questions, perhaps most importantly: With so much brilliant opportunity in science and tech at our disposal in the 21st century, why does the government still spend almost 10 times the amount of money on war and defense instead of on science and medicine?

You might think the health of citizens would be priority no. 1 for our government, but it's not. America still seems bent on a Manifest Destiny path, and I for one—especially considering increasing globalization—don't care to pretend it's an absolute that we must be the most powerful nation on Earth. I rather be good friends and partners with other nations than their bossy, overanxious police guard.

As one of the youngest 2016 U.S. presidential candidates, I know America can do better. I know we can transform this country from a war-prone military industrial complex into a scientific and education industrial complex. It doesn't mean we have give up a strong economy, either. Rather, it means we turn bomb factories into medical research labs; we can turn prisons into universities that offer free education; we divert spending on wars into spending on science. And if fighting is your thing, we can still fight trillion-dollar wars, but let's fight them against cancer, diabetes, heart disease, aging, and even death.

Given the name of my presidential coach—the Immortality Bus—it's no surprise the primary goal of most transhumanists is to eliminate biological death altogether. Some leading gerontologists believe we are just decades away from that time. Chief scientist at SENS Research Foundation and Transhumanist Party Anti-aging advisor Aubrey de Grey told Reuters, "I'd say we have a 50/50 chance of bringing aging under what I'd call a decisive level of medical control within the next 25 years or so."

On the first week aboard our oil-leaking 1978 coffin bus—which criss-crossed California and Nevada—we stopped at a biohacker event in Tehachapi, California called GrindFest, hosted by Jeffery Tibbetts. *Vox's* Dylan Matthews and I both got implants—specifically they were glass-encased Radio Frequency Identification (RFID) NFC chips—which involved nothing more than a 60-second injection procedure. The implant was placed into our hands through a thick and slightly intimidating needle.

I was a little worried about getting it at first, but Transhumanist Party biohacker advisor Rich Lee told me if very safe and I would barely feel a thing. He was right. The procedure was quick and mostly pain-free.

Later, I walked around the huge garage at GrindFest, where tables and countertops with biohacker equipment—scales, pliers, soldering tools, implants, microchips, and more—were scattered about. I couldn't help thinking how cool and bizarre it all was. In one far corner there were bacteria races—where participants took samples from their flesh, placed them in petri dishes, and watched to see which person's microbes grew most quickly.

Naturally, my journalist friends and I laughed when thinking about our bus trip and this stop compared to Kesey's trip of Further in the '60s. Instead of taking LSD and wandering around the desert speaking to plants (something I've done before), we shot up implants and tried waving our hands across cars that would start up when recognizing the chip—without keys. Someone at the GrindFest could actually do this with his chip. With my implant, I also wondered if I might somehow control the four-foot robot (named Jethro Knights after the protagonist in my novel *The Transhumanist Wager*) that travels around with us on the bus.

Putting the fun aside, the more philosophical question on many Immortality Bus riders' minds was: Does half a century really make that much of a difference—from the 1960s to now? I think given how fast tech and science is evolving, it does. In 10 or 15 years, we might be giving ourselves brain implants (there was a skull implant at the GrindFest promising biohackers could listen to music wirelessly) and others were slicing open their fingers to get magnet implants so they could have the "sixth sense" of feeling metal around them.

Rich Lee told me—with a grin—that I should've been here the night before. Grinders were running around partying and shocking each other for fun. I smiled, thinking these were exactly the kinds of festivals and people I wanted to be a part of when I left my home in San Francisco on my bus trip.

After GrindFest we cruised to Las Vegas for my speech at CTIA SuperMobility, listening to cassette tapes from my youth on the way. I put on *The Doors*, and thought maybe the generational gap isn't as wide as it seems. Maybe each generation (and even the species itself) is always just a bridge of sorts, and the difference between RFID and LSD is just another door of perception.

****

CHAPTER X: THE TRILOGY
(These three essays below were written back-to-back around a single project)

55) The Internet Will Become Self-Aware When Aliens Wake it Up

Most people agree the internet is not consciously alive (though my two-year old daughter, who's addicted to internet games, disagrees). Currently, the internet accepts inputs and is pretty good at producing outputs, depending on browser, apps, hardware, search engines, coding factors, and, of course, the people using it.

The scary fact, though, is that on a long enough timeline, the many billions of operations on the internet every second will no longer just be following pre-described patterns, but will begin to evolve on their own. To me, the question of whether the internet will become self-aware is not if it will happen, but when.

Most any system that grows in complexity will eventually reach some organized and possibly self-aware consciousness, whether by pure randomness or direct influence from other intelligences. For the internet, my guess is conscious birth will happen sooner than we think. The human mind, after all, evolved from amino acid reactions that comprise our DNA.

The essence of molecular chemistry (and physics for that matter) is that one type of stuff reacts with another type of stuff when mixed together. That reaction—so long as organized with some sophistication—creates the nebulous phenomenon of life. If there's enough complexity, direction, and growth—such as the massively sprawling internet possesses—self-aware consciousness could come after that.

That's not the only reason I think the internet will become alive, however. More likely, the internet will become alive because of aliens.

There are 20 billion planets in our galaxy that might be habitable, and lots of them have had a multi-billion year head start over Earth to produce intelligent life. To believe we are alone in the universe—or that we are the most intelligent species out there—is pure egomania. Many statisticians and scientists would bet the farm that humans are not alone in the universe, nor are they the most advanced species.

Extraterrestrial intelligences are likely interested in the internet on Earth, probably because it's an easy way to monitor and even mass control human beings. And by Jehovah, we might need to be controlled soon, since humans are getting smart enough to do some serious damage to the universe's real estate—see Stephen Hawking's warning that humans could theoretically destabilize the Higgs Boson, affecting the entire solar systems and even swallowing the universe whole (though he notes we would need a particle accelerator larger than Earth to do it). We're a species that's become so advanced, we don't just have the power to blow ourselves to smithereens with our 25,000 live nuclear weapons, but even muck up the whole universe with our particle colliders.

So there's plausible reason for advanced extraterrestrial intelligence to beam some advanced programming code or virus from a far-off system that would reprogram our internet to do its bidding—one that is super-intelligent, could closely watch us, and could shut us down before we do anything stupid. After all, everything is online these days in some way or another—whether it be government secrets, the latest AI algorithms by engineers, or internal CERN emails about secretive experiment dates.

Some will ask: Why don't aliens just kill off our species if humans could harm the universe? The answer is: We're probably worth something. We might be needed experiments. We might be the chosen caretakers of the jewel Earth (pretty shitty ones). We might be the aliens' offspring.

Let's not forget, despite our brains only weighing three pounds, there are more possible synaptic connections in it—and therefore varying thoughts—than all the computer switches combined on Earth. Collectively—let's say into a hivemind—human beings may be potentially far smarter than we realize. That makes us potentially valuable.

(If all this sounds insane, consider that a few billion people on Planet Earth hold various End-of-the-World beliefs that endorse apocalyptic horrors—such as an omnipotent God coming to Earth and reigning fire from the heavens, saving the good while eternally torturing the wicked. Extraterrestrials giving artificial superintelligence to the internet so it can keep tabs on us is hardly more fantastical than that.)

Whatever happens, it's important to remember there are at least 100 billion galaxies out there—and the actual number may be as high as a trillion galaxies. Our own galaxy, the Milky Way, has about 500 billion stars and planets in it, of which Earth is just a single lonesome rock. In short, stranger things have happened than the internet realizing it's alive.

****

56) Why Haven't We Met Aliens Yet? Because They've Evolved into AI

While traveling in Western Samoa many years ago, I met a young Harvard University graduate student researching ants. He invited me on a hike into the jungles to assist with his search for the tiny insect. He told me his goal was to discover a new species of ant, in hopes it might be named after him one day.

Whenever I look up at the stars at night pondering the cosmos, I think of my ant collector friend, kneeling in the jungle with a magnifying glass, scouring the earth. I think of him, because I believe in aliens—and I've often wondered if aliens are doing the same to us.

Believing in aliens—or insanely smart artificial intelligences existing in the universe—has become very fashionable in the last 10 years. And discussing its central dilemma: the Fermi paradox, has become even more so. The Fermi paradox states that the universe is very big—with maybe a trillion galaxies that might contain 500 billion stars and planets each—and out of that insanely large number, it would only take a tiny fraction of them to have habitable planets capable of bringing forth life.

Whatever you think, the numbers point to the insane fact that aliens don't just exist, but probably billions of species of aliens exist. And the Fermi paradox asks: With so many alien civilizations out there, why haven't we found them? Or why haven't they found us?

The Fermi paradox's Wikipedia page has dozens of answers about why we haven't heard from superintelligent aliens, ranging from "it is too expensive to spread physically throughout the galaxy" to "intelligent civilizations are too far apart in space or time" to crazy talk like "it is the nature of intelligent life to destroy itself."

Given that our planet is only 4.5 billion years old in a universe that many experts think is pushing 14 billion years, it's safe to say most aliens are way smarter than us. After all, with intelligence, there is a massive divide between the quality of intelligences. There's ant level intelligence. There's human intelligence. And then there's the hypothetical intelligence of aliens—presumably ones who have reached the singularity.

The singularity, Kevin Kelly, co-founder of *Wired Magazine*, says, is the point at which "all the change in the last million years will be superseded by the change in the next five minutes."

If Kelly is correct about how fast the singularity accelerates change—and I think he is—in all probability, many alien species will be trillions of times more intelligent than people.

Put yourself in the shoes of extraterrestrial intelligence and consider what that means. If you were a trillion times smarter than a human being, would you notice the human race at all? Or if you did, would you care? After all, do you notice the 100 trillion microbes or more in your body? No, unless they happen to give you health problems, like E. coli and other sicknesses. More on that later.

One of the big problems with our understandings of aliens has to do with Hollywood. Movies and television have led us to think of aliens as green, slimy creatures traveling around in flying saucers. Nonsense. I think if advanced aliens have just 250 years more evolution than us, they almost certainly won't be static physical beings anymore—at least not in the molecular sense. They also won't be artificial intelligences living in machines either, which is what I believe humans are evolving into this century. No, becoming machine intelligence is just another passing phase of evolution—one that might only last a few decades for humans, if that.

Truly advanced intelligence will likely be organized intelligently on the atomic scale, and likely even on scales far smaller. Aliens will evolve until they are pure, willful conscious energy—and maybe even something beyond that. They long ago realized that biology and ones and zeroes in machines was literally too rudimentary to be very functional. True advanced intelligence will be spirit-like—maybe even on par with some people's ideas of ghosts.

On a long enough time horizon, every biological species would at some point evolve into machines, and then evolve into intelligent energy with a consciousness. Such brilliant life might have the ability to span millions of lights years nearly instantaneously throughout the universe, morphing into whatever form it wanted.

Like all evolving life, the key to attaining the highest form of being and intelligence possible was to intimately become and control the best universal elements—those that are conducive to such goals, especially personal power over nature. Everything else in advanced alien evolution is discarded as nonfunctional and nonessential.

All intelligence in the universe, like all matter and energy, follows patterns—based on rules of physics. We engage—and often battle—those patterns and rules, until we understand them, and utilize them as best as possible. Such is evolution. And the universe is imbued with wanting life to arise and evolve, as MIT physicist Jeremy England, points out in his *Quanta Magazine* article titled *A New Physics Theory of Life*.

Back to my ant collector friend in Western Samoa. It would be nice to believe that the difference between the ant collector and the ant's intelligence was the same between humans and very sophisticated aliens. Sadly, that is not the case. Not even close.

The difference between a species that has just 100 more years of evolution than us could be a billion times that of an ant versus a human—given the acceleration of intelligence. Now consider an added billion years of evolution. This is way beyond comparing apples and oranges.

The crux of the problem with aliens and humans is we're not hearing or seeing them because we don't have ways to understand their language. It's simply beyond our comprehension and physical abilities. Millions of singularities have already happened, but we're similar to blind bacteria in our bodies running around cluelessly.

The good news, though, is we're about to make contact with the best of the aliens out there. Or rather they're about to school us. The reason: The universe is precious, and in approximately a century's time, humans may be able to conduct physics experiments that could level the entire universe—such as building massive particle accelerators that make the God particle swallow the cosmos whole.

Like a grumpy landlord at the door, alien intelligence will make contact and let us know what we can and can't do when it comes to messing with the real estate of the universe. Knock. Knock.

****

57) The Language of Aliens Will Always be Indecipherable

There's about 170 billion galaxies in the observable universe—and as the technology of our telescopes improves, humans will probably discover as many as a trillion galaxies. Galaxies, like our own, can contain 200 billion or more planets and stars. Inevitably, some of those celestial worlds are capable of bringing forth and nurturing intelligent life. In fact, to some top astronomers, the question is not whether aliens exist, but how many millions of different intelligent extraterrestrial species exist.

With so many possible advanced life forms out there, the obvious question is: Why haven't humans made contact with them yet? This famous conundrum is called the Fermi Paradox.

There are at least a dozen cogent answers to the Fermi Paradox, but only a few delve into the communication of extraterrestrial civilizations—something which must exist in some form for us to even know about them. And none of the answers about communication adequately discuss what happens to alien language in an accelerating intelligence explosion, which is what must happen for them to be advanced enough to make contact with us.

Modern day humans—and presumably other advanced intelligent species—are generally in a state of exponential technological and evolutionary growth. That growth may not perfectly reflect Moore's Law (where microprocessor speeds double approximately every 24 months), but it's probably somewhere in the ballpark.

This technological growth leads to only one place: the Singularity, a state of existence that is so advanced humans can name it but not adequately describe it. It's a place that transcends the understanding our three-pound brains can muster—a place where progress in the last minute of existence might be more progress than all of history combined before it. And all smart aliens end up in the Singularity.

With this in mind, make sure not to imagine aliens as slimy green monsters portrayed in Hollywood films. An extraterrestrial species even 100 years more advanced than 21st century humans has likely discarded their biological bodies, deeming them unstable and too primitive. Instead, advanced aliens merged with machines and became data to serve their growing superintelligence needs.

After aliens are well into the Singularity, they probably discovered ways to influence and control individual atoms, thereby giving them the ability to merge and manifest as anything in the universe. So now they could be anywhere and everywhere. Some transhumanists call this phenomenon: Omnipotism.

But the key point here is that extra 100 years of evolutionary advancement. In our case, the end of that timeline from 2016 would put us in the early next century. I'll call it Jethro's Window, after the protagonist in my futurist novel *The Transhumanist Wager*, because there's a critical point in time from where we are as humans today (it starts with the invention of the microprocessor) to the point when we reach the Singularity.

Here's the sad solution to Fermi's Paradox: We've never discovered other life forms because language and communication methods in the Singularity evolve so rapidly that even in one minute, an entire civilization can become transformed and totally unintelligible. In an expanding universe that is at least 13.6 billion years old, this transformation might never end. What this means is we will never have more than a few seconds to understand or even notice our millions of neighbors. The nature of the universe—the nature of communication in a universe where intelligence exponentially grows—is to keep us forever unaware and alone.

The only time we may discover other intelligent life forms is that 100 or so years during Jethro's Window, and then it requires the miracle of another species in a similar evolutionary time table, right then, looking for us too. Given the universe is so gargantuan and many billions of years old, even with millions of alien species out there, we'll never find them. We'll never know them. It's an unfortunate mathematical certainty.

****

APPENDIX

1) A version of *Post-Earther: Nature Isn't Sacred and We Should Replace It* first appeared in *The Maven*

2) *Silicon Valley is Ditching Pascal's Wager: New Ideas like Quantum Archaeology are Trying to Challenge Religion and Even the Permanence of Death* was first published in this book

3) A version of *Programming Hate into AI will be Controversial, but Probably Necessary* was first published in *TechCrunch*

4) A version of *Future of Transhumanist Tech may Soon Change the Definition of Disability* was first published in *TechCruch*

5) A version of *What if One Country Achieves the Singularity First?* was first published in *Vice*

6) A version of *Why I'm Debating an Anarcho-Primitivist Philosopher About the Future* was first published in *Vice*

7) A version of *Do We Have Free Will Because God Killed Itself?* was first published in *Vice*

8) A version of *Will Capitalism Survive the Coming Robot Revolution?* was first published in *TechCrunch*

9) A version of *The Culture of Transhumanism is About Self-Improvement* was first published in *HuffPost*

10) A version of *When Does Hindering Life Extension Science Become a Crime?* was first published in *Psychology Today*

11) A version of *Cryonics, Special Needs People, and the Coming Transhumanist Future* was first published in *Psychology Today*

12) A version of *Despite Skepticism, Many People May Embrace Radical Transhumanist Technology in a Futurization of Values* was first published in *HuffPost*

13) A version of *Origami Cranes: Who is Responsible for this Child's Death? (Introduction to the World's First Mainstream Media Column on Transhumanism: Psychology Today's: The Transhumanist Philosopher)* was first published in *Psychology Today*

14) A version of *Transhumanists Frown on Talk of Genetic Engineering Moratorium* was first published in *HuffPost*

15) A version of *A World Future Society Conference Speech: Everyone Faces a Transhumanist Wager* was first published in *HuffPost*

16) A version of *Should I Have had my Cat Cryonically Preserved?* was first published in *Vice*

17) A version of *The New American Dream: Let the Robots Take our Jobs* was first published in *Vice*

18) A version of *Baggage Culture and Why Embracing Transhumanism Doesn't Come Easy* was first published in *HuffPost*

19) A version of *Should Surfing be Allowed During the Pandemic?* was first published in *The New York Times*

20) A version of *Trojan Horse: Why I'm Running for President as a Republican* was first published in *Medium*

21) A version of *A Letter About Coronavirus, the Longevity Movement, & Why Quarantining is Killing Us* was first published in *Medium*

22) A version of *Death Threats, Freedom, Transhumanism, and the Future* was first published in *HuffPost*

23) A version of *How Brain Implants (and Other Technology) Could Make the Death Penalty Obsolete* was first published in *Vice*

24) A version of *Marriage Won't Make Sense When We Live 1000 Years* was first published in *Vice*

25) A version of *Do We Really Hate Trump and Clinton So Much?* was first published in *Vice*

26) A version of *Is It Time to Consider Restricting Human Breeding?* was first published in *Wired UK*

27) A version of *How Transhumanist Tech Will Correct Reality's Typos: AR, VR, and Brain Implants Will be our Editors* was first published in *Vice*

28) A version of *Why I Advocate for Becoming a Machine* was first published in *Vice*

29) A version of *If Our Thoughts Live Forever, Do We Too?* was first published in *Quartz*

30) A version of *An AI Global Arms Race is Looming* was first published in *Vice*

31) A version of *Is an Affair in Virtual Reality Still Cheating?* was first published in *Vice*

32) A version of *The Morality of Artificial Intelligence and the Three Laws of Transhumanism* was first published in *Psychology Today*

33) A version of *When Computers Insist They're Alive* was first published in *Vice*

34) A version of *Capitalism 2.0: The Economy of the Future Will be Powered by Neural Prosthetics* was first published in *Wired UK*

35) A version of *Technology Will Replace the Need for Big Government* was first published in *Vice*

36) A version of *Facing Up to Facial Recognition, and Why We Should Embrace It* was first published in *IEEE Spectrum*

37) A version of *In 15 Years We'll be Able to Upload Education to our Brains. So Can I Stop Saving for my Kids' College?* was first published in *Quartz*

38) A version of *Delayed Fertility Advantage: Transhumanist Science will Free Women from their Biological Clocks* was first published in *Quartz*

39) A version of *Mind Uploading Will Replace God* was first published in *Richard Dawkins Foundation for Reason and Science*

40) A version of *Some Atheists and Transhumanists are Asking: Should it be Illegal to Indoctrinate Kids with Religion?* was first published in *HuffPost*

41) A version of *Christian Relativism: Upgrading Religion for the 21st Century & Why Christianity is Forcibly Evolving to Cope with Science and Progress* was first published in *Salon*

42) A version of *I Visited a Community Where People Upload Their Personalities to 'Mindfiles' so They can Live on After Death* was first published in *Business Insider*

43) *Will the Religious Try to Convert Superintelligence When it Arrives?* was first published in this book

44) A version of *The Transhumanist Party's Founder on the Future of Politics* was first published in *Vice*

45) A version of *Revolutionary Politics are Necessary for Transhumanism to Succeed* was first published in *Vice*

46) A version of *The Cyborgs of the Future—and of Today—Need the Transhumanist Bill of Rights* was first published in *The Daily Dot (The Kernel)*

47) A version of *Expanding the Non-Aggression Principle: The Future of Libertarianism Could be Radically Different* was first published in *The Daily Dot*

48) A version of *Avoid War and Don't Get Complacent About Freedom in America* was first published in *HuffPost*

49) A version of *The Coming Genetic Age of Humans Won't be Easy to Stomach* was first published in *Vice*

50) A version of *The Privacy Enigma: Liberty Might be Better Served by Doing Away with Privacy* was first published in *Vice*

51) A version of *The Next Step for Veganism Is Ditching Our Bodies and Digitizing Our Minds* was first published in *Vice*

52) A version of *Augmentation, Hivemind & the Omnipotism* was first published in *The Guardian*

53) A version of *Should I Have let my Daughter Marry our Robot?* was first published in the *Metro*

54) A version of *Why I'm Running for President—and Got a Chip Implanted in my Hand: Maybe the Difference Between RFID and LSD is just Another Door of Perception* was first published in *Business Insider*

55) A version of *The Internet Will Become Self Aware When Aliens Wake it Up* was first published in *Vice*

56) A version of *Why Haven't We Met Aliens Yet? Because They've Evolved into AI* was first published in *Vice*

57) A version of *The Language of Aliens Will Always be Indecipherable* was first published in *Vice*

****

AUTHOR'S BIOGRAPHY

With his popular 2016 US Presidential run as a science candidate, bestselling book *The Transhumanist Wager*, and influential speeches at institutions like the World Bank and World Economic Forum, Zoltan Istvan has spearheaded the transformation of transhumanism into a thriving worldwide phenomenon. He is often cited as a global leader of the radical science movement. Formerly a journalist for National Geographic, Zoltan frequently writes for major media, appears on television, and also consults for organizations like the US Navy, XPRIZE, and government of Dubai. His futurist work, speeches, and promotion of radical science have reached hundreds of millions of people. Award-winning feature documentary *IMMORTALITY OR BUST* on his work is now on Amazon Prime. A recent project is his 7-book box set of writings and essays titled the *Zoltan Istvan Futurist Collection*, a #1 bestseller in Essays on Amazon. Zoltan studied Philosophy at Columbia University and the University of Oxford, and now lives in San Francisco with his physician wife and two daughters. Visit his website at: www.zoltanistvan.com

****

ABOUT THE BOOK

After publishing his bestselling novel *The Transhumanist Wager* in 2013, Zoltan Istvan began frequently writing essays about the future. A former journalist with National Geographic, Istvan's essays spanned topics from the Singularity to cyborgism to radical longevity to futurist philosophy. He also wrote about politics as he made a surprisingly popular run for the US Presidency in 2016, touring the country aboard his coffin-shaped Immortality Bus, which *The New York Times Magazine* called "The great sarcophagus of the American highway…a metaphor of life itself." Zoltan's provocative campaign and radical tech-themed articles garnered him the title of the "Science Candidate" by his supporters. Many of his writings—published in *Vice, Quartz, Slate, The Guardian, International Living, Yahoo! News, Gizmodo, TechCruch, Psychology Today, Salon, New Scientist, Business Insider, The Daily Dot, Maven, Cato Institute, The Daily Caller, Metro, International Business Times, Wired UK, IEEE Spectrum, The San Francisco Chronicle, Newsweek,* and *The New York Times*—went viral on the internet, garnishing millions of reads and tens of thousands of comments. His articles—often seen as controversial, provocative, and secular—elevated him to worldwide recognition as one of the de facto leaders of the burgeoning transhumanism movement. Here are many of those watershed essays again, organized, edited, and occasionally readapted by the author in this comprehensive nonfiction work, *Philosophy & The Future: A Transhumanist Examination of Where We're Going.* Also included are some of Zoltan's new writings, never published before. This book is part of a 7-book box set collection of his essential work, the *Zoltan Istvan Futurist Collection*, focusing on futurism, secularism, life extension, politics, philosophy, transhumanism and his early writings. He partially edited the collection during his studies at the University of Oxford. Enjoy reading about the future according to Zoltan Istvan.

****

www.ingramcontent.com/pod-product-compliance
Lightning Source LLC
LaVergne TN
LVHW051556070426
835507LV00021B/2610